An Introduction to Freshwater Ecology

D. H. Mills

M.Sc., Ph.D.

Lecturer in Freshwater Ecology
and Fisheries Management
Department of Forestry and Natural Resources
University of Edinburgh

Oliver & Boyd · Edinburgh

73931

574
.929
MIL

OLIVER & BOYD
Tweeddale Court
14 High Street
Edinburgh EH1 1YL
A Division of Longman Group Limited

ISBN   0 05 002397 7   0053591

Printed in Great Britain by
Cox & Wyman Limited
London, Fakenham and Reading

An Introduction
to Freshwater Ecology

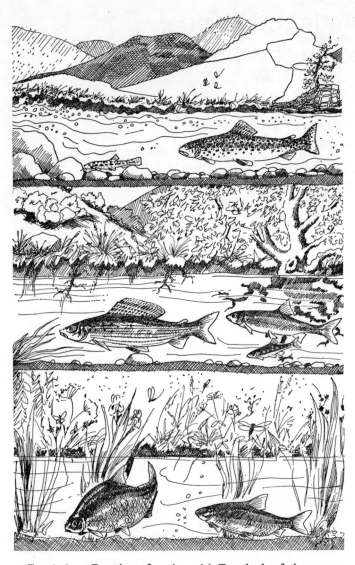

Frontispiece. Zonation of a river. (*a*) Troutbeck of the upper mountainous course (Stone loach and trout); (*b*) Grayling zone or minnow reach of the foothills (Grayling, dace and minnow); (*c*) Coarse fish reach of the lowland plains (Roach). (See Chapter 3).

(Artist: Christopher Lee)

Contents

Preface

Water has a tremendous fascination for most of us. We like to swim in it, walk by it or sail on it. There are many hobbies and pursuits which depend on it – sailing, fishing, canoeing, swimming and water skiing. Whatever our interest is we are usually keen to know more about the life under the water surface or by the water's edge; the fisherman wants to know what fish he is likely to catch and what they are eating, the canoeist and yachtsman may be more interested in the water weeds, while the rambler is attracted by the water birds and waterside plants.

Let us take a look at the life in the freshwater around us. Some of you will be able to look at streams, rivers or lakes, while others will be in a better position for studying canals and ponds or disused gravel-pits.

The type of study we are going to do is called ECOLOGY – a study of the inter-relations between living organisms and their environment. This study can be divided into *autecology* which involves the correlation of the distribution of an animal with environmental factors, and *synecology* which deals with the biological relationships of populations within a community. Both aspects are considered in this text.

The suggested exercises at the end of each chapter play an important part and are designed both to help with field studies and projects, and as encouragement to delve more deeply into this fascinating subject.

Acknowledgements

I should like to thank many of my students whose work in the field has produced some of the results which appear in this book, and whose enthusiasm has been a great stimulus to my own teaching. I greatly appreciate the work of the artist, Christopher Lee, whose work appears in the Frontispiece and in Figs. 5.1 and 5.2.

My acknowledgements are also due to the following for kindly allowing me to reproduce some of this material: Edward Arnold Ltd. (Fig. 6.3); Association of River Authorities (Fig. 8.7); Blackwell Scientific Publications Ltd. (Figs. 6.8, 9.1); E. J. Brill (Fig. 8.8); Collins Ltd. (Table 1); Controller of Her Majesty's Stationery Office (Figs. 4.1, 8.4 and Table 13); Dr H. J. Egglishaw (Fig. 5.4 and Table 2); Fishing News (Books) Ltd. (Table 8); Dr R. B. Herrington (Tables 5 and 6); Holt, Rinehart and Winston Inc. (Figs. 6.5, 7.2 and 7.3); Professor M. Huet (Table 11); International Biological Programme (Fig. 9.1); Mr E. D. Le Cren (Figs. 6.8, 8.7); Liverpool University Press (Fig. 10.1); Longman (Figs. 1.2, 5.3, 5.5 and Table 9); Macmillan Ltd. (Figs. 1.4, 6.1 and 6.2); McGraw-Hill (Fig. 2.1 and Table 3); Dr K. H. Mann (Figs. 6.6, 6.7); Mr I. Mitchell (Figs. 1.1, 1.4 and 4.2); Mr N. C. Morgan (Table 2); Prentice-Hall Ltd. (Fig. 8.6); W. B. Saunders Company (Fig. 6.4); Trent River Authority (Table 12); U.S. Department of Agriculture (Fig. 7.1).

Edinburgh, 1972   DEREK MILLS

1 Physical and Chemical Characteristics of Water

Before we look at the various freshwater habitats and the plants and animals that live in them, we must know something about the properties of freshwater. The chemical and physical characteristics or factors are most important in that they play a great part in influencing the distribution and behaviour of plants and animals.

Physical

Buoyancy Water is denser than air, so it is a much more buoyant medium. It exerts an upthrust on all plants and animals which therefore do not have to support their own weight. This buoyancy enables aquatic plants and animals to have a much more delicate structure than terrestrial ones, and also enables animals to live at various depths.

Temperature Water has the greatest specific heat of all substances and so can absorb a relatively large amount of heat for a small rise in its temperature. It therefore acts as a buffer against wide fluctuations in temperature and so lakes and rivers change their temperature relatively slowly (Fig. 1.1). Warm water is less dense than cold water which is densest at 4 °C. In the autumn, when the surface of a lake cools down the upper layers of the water sink and displace warmer water from below. This process goes on until the temperature is uniform from top to bottom. Water colder than 4 °C is less dense because of a change in the structure of the water molecules and therefore floats at the surface, where further cooling leads to the formation of ice.

The range of many plants and animals is limited by water temperature, and the growth rate of animals is also to a great extent controlled by temperature, because various metabolic processes governing growth react to critical temperatures.

Pressure Water is almost incompressible and there is therefore not a big increase in density with increase in depth. Anything with a specific gravity higher than that of water will go on sinking until it

1

reaches the bottom of the lake or stream. For every 10 metres that an object sinks below the surface of the water the pressure upon it increases by one atmosphere. The pressure inside an aquatic animal is almost the same as the pressure outside.

Viscosity   The viscosity or internal friction of water varies with temperature. Water is twice as viscous near freezing-point than at ordinary summer temperature and this affects the rate at which small bodies sink, and the ability of small animals to maintain their position in the water.

Surface tension   This is an important physical factor for small plants and animals. Water crickets and water striders can support themselves on the surface of the water by it, and snails and flatworms can crawl along the underside of the surface film. Land insects which fall on the water surface are often trapped in the surface film because of this surface tension.

Light   As water reflects more light than the land a certain amount of light is lost to submerged plants at the water surface. Furthermore, light and its energy diminish with distance from the surface until there may be virtually none below a certain depth. Because of the different absorption of wave-lengths by water the quality of light changes with depth. Water readily absorbs the shortest wave-lengths of solar energy and within the range of visible light it absorbs the longer wave-lengths more effectively than it does the shorter. So the light that penetrates to the deeper parts of a lake, for example, is relatively poor in red and orange rays, which are the most effective for photosynthesis, but relatively rich in green and blue rays as compared with light at the surface of the lake. Because of this differential absorption of light components by water, plants in the deeper part of the illuminated zone do not receive the same sort of light as do those in the upper part.

Conductivity   The electrical conductivity of water depends, almost entirely, upon the substances dissolved in it (the *electrolytes*); the conductivity of a water being proportional to the amount of its dissolved substances. The unit of conductivity normally used is the reciprocal of the megohm and is expressed as reciprocal megohms. It is common practice to express reciprocal megohms as micromhos or μmhos. Conductivity is then written in terms of micromhos/cm or μmhos/cm. As temperature influences the reading, a correction to 20 °C is made to ensure uniformity. Distilled water has a conductivity of practically zero while waters running over rock rich in soluble salts have a high conductivity, for example the chalk streams in southern England may have a value of 400 to 450 reciprocal megohms. Moorland streams running over mineral-poor rock, on the other hand, have conductivities in the region of 100 or less, streams and lochs in the Scottish Highlands may have values as low as 30.

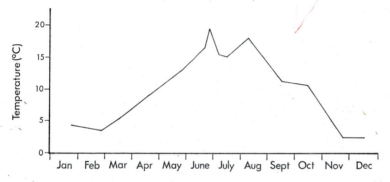

Fig. 1.1. Seasonal changes in water temperature at 100 cm depth in Duddingston Loch, Midlothian, 1970 (from: Mitchell, 1971). (The low water temperature in July reflects the cold, wintry weather of late June and early July.)

Chemical

Oxygen　One of the most important chemical factors in a water environment is the amount of oxygen in the water, because nearly all plants and animals require this gas to breathe. The water temperature governs the amount of oxygen that can be dissolved in the water. It can be stated that:

> The amount of oxygen that a given volume of water will hold in equilibrium drops as temperature rises, and the concentration can be expressed as the percentage of what it would be if the water were saturated at normal pressure and the temperature in question (Fig. 1.2).

Under average conditions at 0 °C there will be about 100 parts per million of oxygen in one litre of water. The concentration falls with rising temperature and at 20 °C there will only be about 65 parts of oxygen. Still water with much vegetation in bright sunlight may, for a period, have more oxygen in solution than the normal maximum at the temperature prevailing. This fluctuation in oxygen concentration is noticeable in ponds and shallow lakes and is called an *oxygen pulse*. Photosynthetic activity during the day results in a maximum oxygen concentration before dusk, but during the night respiration reduces the available oxygen to a minimum before dawn. The oxygen concentration of the water is also affected by water turbulence and decomposition, as well as photosynthesis and respiration.

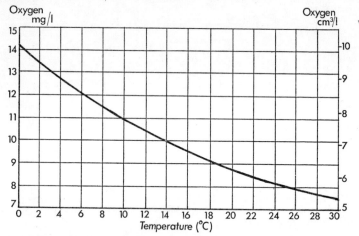

Fig. 1.2. The oxygen content of water saturated with air at normal pressure (760 mm/Hg) (Mortimer 1956) (Mitt. int. Limnol. No. 6, fig. 1) (from Macan).

Carbon dioxide The decomposition of organic matter and the respiration of plants and animals produce carbon dioxide. This gas combines chemically with water to produce carbonic acid which affects the hydrogen ion concentration or pH of the water. Carbonic acid (H_2CO_3) dissociates to produce hydrogen ($H+$) and bicarbonate (HCO_3^-) ions. The bicarbonate radical may undergo further dissociation forming more hydrogen ($H+$) and carbonate (CO_3^{--}). The amount of free carbon dioxide in water is about 0·5 ml of carbon dioxide per litre of water, but much more carbon dioxide is present in ionized form as bicarbonate (HCO_3^-) and carbonate (CO_3^{--}) radicals. When carbonic acid dissociates it releases hydronium ions (H_3O) which alters the pH of the water. At low pH values, when the water is acid (*i.e.* with a pH below 7·0), most carbon dioxide is present in free form, but at or close to neutrality (i.e. a pH of 7·0) most carbon dioxide exists as HCO_3^-, and at high pH values most is present as

CO_3^{--}. So it can be seen that the anions HCO_3^- and CO_3^{--} increase as conditions become more alkaline, thus tending to resist hydrogen ion changes. This is called *buffer action* or buffering which effectively reduces hydrogen ion variability in ponds, lakes and streams. The alkalinity of a water can therefore be defined as the amount of bicarbonate and carbonate ions in solution. It is limited by the availability of calcium and magnesium. Hydrogen ion or pH values in fresh waters may vary from 2·0 to 12·0 but the majority of fresh waters have a pH ranging between 5·5 and 10·0. In acid or poorly buffered waters with abundant vegetation the photosynthetic activity of plants reduces the amount of carbon dioxide during daylight, but at night the respiratory activity of plants and animals leads to an increase in carbon dioxide. These changes in carbon dioxide content lead to changes in pH (Fig. 1.3), causing what is known as a *pH pulse*. A similar pulse is not as detectable in alkaline waters, as the buffering action reduces pH variation to a minimum (Fig. 1.4).

The pH can be defined as the logarithm of the reciprocal of the hydrogen ion (i.e. the hydronium ion) activity. It can be expressed mathematically as: $pH = \log \dfrac{1}{(H^+)}$ where (H^+) is the amount of hydrogen ions in a solution in moles per litre. The amount of free or uncombined carbon dioxide in water is of ecological importance as it governs the precipitation of calcium in the form of calcium carbonate. Acidity impedes the recirculation of nutrients by reducing the rate of decomposition and inhibiting nitrogen fixation. In the presence of calcium carbonate colloidal humus gels are flocculated and the mean particle size of the soil is increased, thereby improving permeability and aeration. At the same time, the ability of these colloids to liberate hydrogen by interacting with neutral salts is reduced, resulting in conditions more favourable for rapid bacterial decomposition of organic matter accumulating in the bottom.

Dissolved substances Water is an extremely good solvent and more chemical compounds will dissolve in water than in any other natural liquid. The quality and quantity of these dissolved substances depends on the geology of the land over which the water has flowed. The main substances in solution are chlorides, sulphates, carbonates, phosphates and nitrates usually in combination with sodium, potassium, magnesium and calcium. Soft or acid waters have very low concentrations of calcium, magnesium, carbonate and sulphate while hard or alkaline waters have high concentrations of these

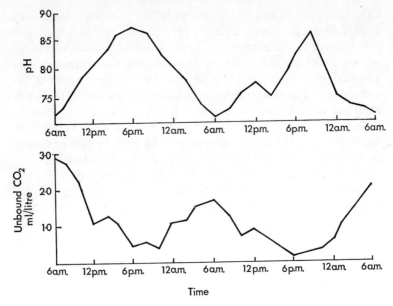

Fig. 1.3. Changes in the hydrogen ion concentration of surface water in a freshwater pond over a two-day period during late summer in the upper graph. The lower graph indicates the changes in carbon dioxide concentration which increases during the dark portion of the forty-eight-hour interval owing to an interruption of photosynthetic activity. (from: Knight, 1965).

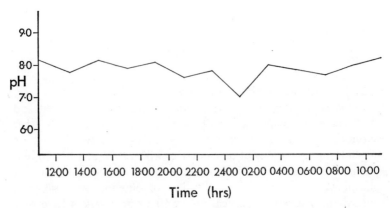

Fig. 1.4. Changes in pH at 100 cm depth over a 24 hour period, Duddingston Loch, Midlothian (from: Mitchell, 1971).

substances. Soft waters are found in drainage areas having an acid soil and are characteristic of many waters in northern England and the Highlands of Scotland. Hard waters are found in areas containing limestone and chalk. The water of south-east England is hard. (Hard water needs a lot of soap to produce a lather while soft water lathers easily.) Very soft water contains humic acids and these may be responsible for the corrosion of pipes. Some of the main substances in solution in some of the chief types of water have been listed by Macan and Worthington (1951) (Table 1). Ennerdale is given as an extreme example of a soft water while Cambridge tapwater is shown as a fairly typical hard water derived from a drainage area in which there are chalk downs. The radicles present in much greater amounts in the Cambridge water than the Ennerdale water are calcium, magnesium and carbonate. Cambridge tapwater now contains less calcium than is shown in Table 1 as a water softener was installed by the Cambridge Water Supply Company. The Burton well-water has an unusually large number of radicles present in fairly high concentrations. The permanent hardness of the water is due to the calcium sulphate or *gypsum*. The reason for the high concentration of sodium in the Braintree water is that the water, originally containing calcium carbonate due to its draining from a calcareous region, passes through the Thanet Sands which are marine in origin. The sodium in the sand displaces the calcium.

TABLE 1

The metallic and acidic radicles of the commoner dissolved substances in certain certain natural waters: figures in parts per million.

	Ennerdale Cumberland	Cambridge tap-water	Braintree Essex	Burton Well	Sea Water	The Dead Sea
Sodium	5·8	1·8	343	51	10 720	142 945
Potassium		7·5		57	380	44 186
Magnesium	0·72	26·5	15	39	1 320	413 241
Calcium	0·8	51	17·5	159	420	171 534
Chloride	5·7	18	407	90	19 320	1 819 600
Sulphate	4·6	16	96	378	2 700	6 237
Carbonate	1·2	123	218	121	70	trace
Silicon dioxide	1·6	—	2	9	1·4	trace

(From: Macan and Worthington, 1951. Reproduced by permission of Collins.)

TABLE 2

The range in certain chemical properties of water samples from streams on different types of rock.

	Number of streams	Total cations	Ca^{++}	Mg^{++}	HCO_3^-	pH
Granite	8	224– 375	40– 95	50–100	26– 100	6·12–6·75
	5	412–1005	138– 361	74–210	100– 296	6·60–7·23
Schist	5	266– 374	56– 115	53– 89	54– 136	6·22–6·75
	11	424–1662	84– 936	29–364	20– 828	5·98–7·40
Basalt	6	730–1061	180– 476	171–315	330– 644	7·13–7·50
Limestone	7	415–1785	195–1457	44–502	100–1108	6·48–7·82
Sandstone	5	1383–3608	656–1063	206–790	328–1232	7·12–7·61

(From: Egglishaw and Morgan, 1965)

Lakes and ponds near the sea usually have a high concentration of chlorides due to the deposition of salt by the sea spray. High chloride concentrations in inland waters may be an indication of pollution.

Egglishaw and Morgan (1965) give the ranges of total cation concentration (as micro-equivalent/1) and certain ions and pH of water samples from Scottish Highland streams occurring on different kinds of rock (Table 2). Only on granite and schist were there streams with water having less than 400 micro-equivalents of total cations/1 and these streams have been classified separately. The extensive range of the total cation concentration/1 of samples from the streams on sandstone is due to three streams which had a high concentration of sodium (644, 1140 and 1705 μe $Na^+/1$), presumably because of their proximity to the sea.

While nitrogen and phosphorus, together with carbon and hydrogen, are the most important constituents of living water, they only occur in freshwater as trace elements. However, they play an important role in the biogeochemical cycle and will be dealt with in Chapter 6.

Suggested Exercises

1. Find out about the geology of your area. The geology of any area is given on geological maps produced by the Geological Survey of Great Britain.

2. Keep a record of the water temperature of a pond or stream for a year and plot the temperature variations on graph paper. Water temperature should be taken near the bed of a stream and at the same place each time. In a lake the temperature should be taken at the same depth each time; surface temperature should also be noted. Reversing thermometers are used for recording water temperature at different water levels. To practise data handling, you can test for differences in mean temperature between months and seasons, using either an analysis of variance or, depending on the sample size, the 't' or 'd' statistic.

3. Keep a record of the air temperature near the above pond or stream and incorporate the results on the above graph. Investigate the relationship by calculating the correlation coefficient between the two temperatures. Calculate the regression equation and try to predict water temperature from the air temperature.

4. Collect water samples from ponds, lakes or rivers in your area and measure the dissolved oxygen, pH and alkalinity. Dissolved oxygen can be estimated by the Winkler method; pH can be recorded with a BDH Comparator or a Mackereth oxygen electrode; a pH meter is a more sophisticated and reliable apparatus. Alkalinity can be estimated by titrating with N/10 sulphuric acid using screened methyl orange as an indicator. The alkalinity (expressed in milligrammes of calcium carbonate per litre) is the volume of sulphuric acid (in cm^3) required for the solution to reach the 'end point' multiplied by 50.

5. Relate your findings to the geology of the area and outside sources such as agriculture and industry.

2 Ponds and Lakes

Freshwater habitats can be divided broadly into two groups; (a) standing water, such as lakes and ponds and (b) running water, such as streams and rivers. Let us first look at some examples of standing water.

A lake can be defined as any large sheet of standing water occupying a basin, while a pond is usually considered to be a small, still body of standing water with rooted plants growing across most of it.

A lake consists of: (i) a littoral or shallow inshore area which extends out to a depth of about 3 metres. This is the area in which rooted plants can grow and is the richest part of the lake. The plants anchor and stabilize the bottom, are a shelter for animals and serve as a surface upon which snails and insects can lay their eggs. The plants also reduce the action of waves against the shore, (ii) an off-shore or limnetic zone of open water in which there are no rooted plants and (iii) a deep water or profundal zone lying beneath the limnetic zone and to which the sun does not penetrate. Not all lakes have this last zone.

Lakes may originate in a number of ways depending on how the lake basin is formed. Lake basins may be formed as a result of movements of the earth's crust, by volcanic or glacial activity, or by stream action, landslides, and even by the impact of meteorites against the earth's surface. Some lakes and lake basins may result from the activities of beavers and aquatic and marsh plants which cause the blocking of valleys with organic materials and logs. Other lakes may be made by man and are known as man-made lakes or reservoirs and are formed by the construction of a dam across the path of a river where it runs through a steep-sided valley. These are usually formed for hydro-electric power, flood control or water storage.

Broadly speaking, the shape of a lake depends upon the way in which the lake basin was formed, for example circular basins are those with a meteoritic or volcanic origin, sub-circular basins (often called tarns) occur in mountainous areas and have a glacial origin,

roughly rectangular lakes have had their basins formed by movements in the earth's crust and crescent-shaped basins hold ox-bow
lakes formed by stream or river action. Not all lakes fit into these
categories and many external factors can alter the shape of a basin,
such as dam and waterway construction and changes in the original
drainage patterns.

There are a number of conditions which determine the nature of
lake bottoms. These include age, size and type of underlying rock and
climate. Young lakes have rocky or sandy bottoms with little deposition of sediments and organic materials, since the sediments accumu-
late as the lake ages. The degree of wave action determines the
amount of eroded material which will be deposited on the bottom
and the structure of the rock formation will determine how much
material is eroded by wave action, rain and wind.

The bottom deposits, which may originate from a number of
sources, include silt deposited by inflowing streams and surface
drainage, eroded material from the shoreline, wind-blown material
and plant and animal remains.

During the summer months many lakes, particularly those that are
deep, may be divided into two layers due to the warming of the
surface water by the sun during calm weather. This layering is called
thermal stratification. The upper layer of warm water is known as the
epilimnion and the lower heavy cold layer is called the *hypolimnion*.
The layer between the epilimnion and the hypolimnion, in which the
fall in temperature is rapid, is called the *thermocline* (Fig. 2.1). Since
light does not penetrate far into water, plant growth is only possible
in the upper layers, and is nearly always confined to the epilimnion.
Algae are present all through the winter. Early in the year these
algae begin to multiply and become very abundant, producing concentrations known as 'blooms'. Shortly after the main production of
algae comes the period when the minute floating animals or zooplankton appear in large numbers. Once a thermocline is established, the
water in the hypolimnion cannot be replenished with oxygen and
consequently the oxygen concentration may fall steadily over the
summer due to the decomposition of plant and animal remains by
bacteria.

It is only the deeper lakes in the British Isles which have a permanent thermal stratification during the summer months. Most of
the shallower lakes show only a very temporary stratification which
is usually broken down at the onset of the first strong wind. With the

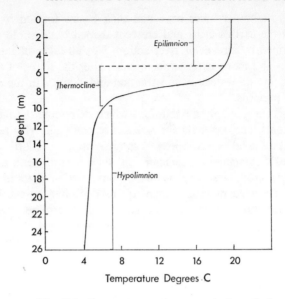

Fig. 2.1. Curve representing a typical vertical distribution of temperature during the summer stagnation period in a temperate lake. The form of the curve is characteristic of the usual condition in thermal stratification. (from: Welch, 1952).

shortening of the days in the autumn and colder air temperatures the epilimnion starts to cool and is eventually obliterated by gale-force winds.

Where layering is well-established a strong wind may cause a *seiche*. This is caused when the wind blows the epilimnion bodily to one side, for the difference in density between the epilimnion and hypolimnion is slight, and pushes it into a deep wedge at the windward end of the lake. This brings the hypolimnion close to the surface at the leeward end of the lake. When the wind drops, the epilimnion piled up on one side rolls back until it is level and then overshoots and piles up on the other side. It continues to oscillate for some time before coming to rest.

Lakes are often classified according to their nutrient content. Those with a low nutrient level are called *oligotrophic* (meaning few foods) and those with a high nutrient content are called *eutrophic* (good foods) lakes.

Oligotrophic lakes are usually deep, have a small littoral zone and lie on infertile rock such as granite. They are poor in dissolved nutrients such as phosphorus, nitrogen and calcium. Oligotrophic lakes are normally situated in mountainous areas and most of the lochs in the Scottish Highlands are oligotrophic. Because of the low production of plants and animals oxygen depletion does not occur in the hypolimnion of this type of lake, which is inhabited by cold water species of fish, such as the brown trout (*Salmo trutta*), char (*Salvelinus alpinus*), whitefish (*Coregonus*) and, in North America, the lake trout (*Cristivomer namaycush*).

Eutrophic lakes lie on mineral-rich rock, such as limestone, and are shallow with a large littoral area rich in flora and fauna. Eutrophic lakes are usually situated on low-lying farmland. Oxygen may become depleted in the hypolimnion during the summer. Although trout may be present in eutrophic lakes, the conditions are generally not satisfactory for the completion of their life-cycle. However, the conditions are suitable for species of fish belonging to the carp family which can tolerate low oxygen concentrations and lay their eggs on vegetation. The members of the carp family most often present are the roach (*Rutilus rutilus*), the bream (*Abramis brama*) and, of course, the carp (*Cyprinus carpio*). *Mesotrophic* lakes have characteristics lying somewhere between those of oligotrophic and eutrophic lakes.

A fourth type of lake is the *dystrophic* lake. This is found mainly in moorland, boggy and mountainous areas. Such lakes are rich in organic matter consisting of undecomposed peat and other humic material. The waters are often stained brown and the pH of the water is usually very low. There is little decomposition because of a deficiency of calcium. This results in a large accumulation of organic matter but a scarcity of nutrients in solution.

A lake gradually ages as sediments and organic material are deposited and fill up the lake basin. This ageing results in deep, steep-sided oligotrophic lakes gradually evolving into shallow, gently shelving eutrophic waters. The rate at which a lake ages depends on the fertility of its drainage basin, the form of the lake basin and the amount of sediment deposited. This ageing process, resulting in enrichment of the water, is known as *eutrophication*. In recent years man has hastened this process by the discharge of nutrients in the form of industrial and domestic waste (as in the Great Lakes of North America) and by intensive farming with the heavy use of fertilizers (as in Loch Leven, Kinross-shire). An interesting example of eutrophication caused by man is that which occurred in the Zürichsee, a lake in the foothills of the Alps. The lake is composed of two distinct basins, the Obersee and the Untersee, separated only by a narrow passage. In the past five decades the deeper of the two, at one time a decidedly oligotrophic lake, became strongly eutrophic owing to urban effluents from a group of small communities. The shallower of the two received no major urban effluent and retained its oligotrophic characteristics. It was noticed that along with domestic fertilization the Zürichsee changed from a whitefish lake to a lake containing cyprinids.

The change in fish populations due to sewage-induced eutrophication has also been shown in the two lake basins of the Bodensee (Lake Constance) on the Swiss–German border. Here the proportions of different fish changed over the period 1910 to 1953. The whitefish, trout, pike–perch and eels formed more than 80 per cent of the 1910–14 catch in one of the basins (the Obersee) and 53 per cent in the other (the Untersee). The percentages in the period 1949–53 were 65 per cent in the Obersee and 38 per cent in the Untersee. During the same period the catch of perch and cyprinids increased in both lakes, particularly in the Obersee. The yield of fish in the Obersee in the late period was close to the early period in the Untersee, which suggests that the sequence for the two lakes may be spliced to give a continuous picture of the change in the species composition in the transition from most oligotrophic (early years in Obersee) to most eutrophic (late years in the Untersee). The process of ageing is an example of what is known as *ecological succession*. This can be defined as: 'The orderly process of community change', that is the sequence of communities which replace one another in a given area, and will be discussed more fully in a later chapter.

Suggested exercises

1. Select a pond or lake for study and note the extent of the littoral and sub-littoral zones, its average and maximum depth and the position of inflowing and outflowing streams. Construct a map of the site and note these features. Shade in areas of stones, gravel and silt, etc.

For those who live in Scotland there is a great deal of physical, chemical and biological information on freshwater lochs in *The Bathymetrical Survey of the Freshwater Lochs of Scotland* by Sir John Murray and Laurence Pullar (published in Edinburgh in 1910 by the Challenger Office). Altogether 562 lochs were surveyed and the results are given in 6 volumes, two of text and four of maps. Volumes 2 to 6 are rather scarce and may be available only in a reference library. However, volume 1 is a most valuable source of information. Information on English Lakes is given in *Biological Studies of the English Lakes* by Macan (1970) and Mill, H. R. (1895) in *Bathymetrical survey of the English lakes: Geographical Journal,* 6, 47–73, 135–166.

The area of lakes (in hectares) can be found on 1:10000 Ordnance Survey maps. Area can also be determined by the use of a planimeter operated on a well-drawn map or aerial photograph of known scale.

The length of a lake is the distance between the furthest points on its shores.

Breadth is the distance from shore to shore measured at right angles to the longitudinal axis. The mean breadth $(\bar{b})$ is obtained by dividing the area (A) by the length, or $\bar{b} = \dfrac{A}{1}$

The mean depth $(\bar{d})$ is a relationship between volume (V) and area (A) of a lake. It can be determined using the formula

$$\bar{d} = \frac{V}{A}$$

However, volume is not easy to determine when mean depth is not known and requires the use of a bathymetric map and the use of the formula

$$V = \tfrac{1}{3}(A_1 + A_2 + \sqrt{A_1 A_2})h$$

where A_1 is the area of the upper surface of a contour stratum and A_2 is the lower surface of the same stratum. The height of the stratum is given by h. The volume for each stratum is calculated by the above formula and the sum of the volumes is the total lake volume.

Volume can be estimated, when mean depth is known, by the formula

$$V = 325\,850A \times d$$
where A = area of lake in hectares
d = mean depth in metres

It is useful to measure the extent and development of a lake shore-line (*SD*). This can be calculated from the formula

$$SD = \frac{S}{2\sqrt{\pi A}}$$

where S = the length of the shore and
A = area of the lake

A value of 1 indicates no development while higher values indicate increasing irregularity of the shoreline.

The main lake-bottom deposits can be classified according to the classification used by Welch (1952):

TABLE 3

I. Homogeneous: having uniform composition.

   A. Inorganic.
     1. Bed rock or solid rock.
     2. Boulders: rocks more than 300 mm in diameter.
     3. Rubble: rocks 50 mm to 300 mm in diameter.
     4. Gravel: 3 mm to 50 mm in diameter.
     5. Sand: may be divided into coarse and fine.
     6. Clay: very finely divided mineral matter; no gritty feeling; usually grey in colour.
     7. Marl: Calcium carbonate; usually grey in colour.

   B. Organic.
     1. Detritus: coarse plant materials, fragmented but little decayed.
     2. Fibrous peat: partially decayed plant remains; parts of plants recogniz-able.
     3. Pulpy peat: very finely divided plant remains; particles unrecognizable; green to brown; consistency variable, often semifluid.
     4. Muck: black; finely divided organic matter; decomposition very advanced.

II. Heterogeneous: composed of two or more kinds of material.
     1. Alluvium: mixed sedimentary material from inflowing streams.
     2. Various combinations of two or more recognizable homogeneous types.

---

2. Try and classify some of the lakes of the world according to how they have been formed. Include as many man-made lakes as you can (*e.g.* Kariba and Volta). A very good account of the charac-teristics and distribution of the lakes of the world is given in Vol. 1 of *The Bathymetrical Survey of the Freshwater Lochs of Scotland.* by Murray and Pullar, 1910.

3. Make a list of the lakes in your area and try to decide whether they are oligotrophic or eutrophic, giving reasons.

4. For those who would like to construct an aquarium model of a lake to demonstrate thermal stratification details are given in an article entitled *A simplified model of a lake for instructional use* by J. R. Vallentyne, published in the *Journal of the Fisheries Research Board of Canada*, Vol. 24, no. 11, 1967.

3 Rivers and Streams

The basic function of rivers and streams (referred to as a *lotic* environment) is to convey the surplus rain water from land to sea. They differ from lakes (a *lentic* environment) in a number of ways, for instance they have:

(*i*) a continuous one-directional movement, with the whole volume of water flowing in one direction. In the longer rivers the flow may be from one climatic zone to another.

(*ii*) a variation in velocity with change in volume of water.

(*iii*) a wide range of fluctuations in water level.

(*iv*) as a rule, a smaller depth compared with lakes.

(*v*) the water is confined to a relatively narrow channel. The only exceptions here are expansions in the river channel sometimes known as river lakes.

(*vi*) physical, chemical and biological conditions gradually changing with distance along the main channel and in a definite direction.

(*vii*) an increase in their length, width and depth with increasing age.

(*viii*) a permanent removal of eroded and transported materials. any materials eroded at one point on a river are transported downstream with no opportunity for return.

(*ix*) an absence of prolonged stagnation.

(*x*) a greater dependence upon the contribution of food materials from the surrounding land, manufacturing little basic food material themselves.

The most significant fact is that rivers and streams are open systems whereas ponds and lakes are closed or self-contained systems except for some gains or losses from the inflowing or outflowing streams. The important point is that in a lake nutrient materials may be used several times while at any point in a river use must be made by plants and animals of material that is there only temporarily.

A river can be classified in two ways; firstly, by its physical characteristics and secondly by the fish species present which indicate the differing physical, chemical and biological features of the river.

The first method divides the river into three parts.

The upper or mountain course. Here the fast-flowing water, particularly after rain, is able to move large stones and roll them along. Angular stones washed into the river are rubbed against one another to form rounded pebbles. Here the river runs through a V-shaped valley with unstable banks and has great powers of erosion.

The middle course of the river occurs over the foothill belt where the water velocity is less and it flows a little more slowly, but still fast enough to carry sand, silt and mud in suspension and to roll pebbles along its bed. Here the main work of the river is transportation of silt. In this part of the river course the valley is broad and has stable sides so that the river does not erode the land as much as it does in the mountain course.

The lower course is that in which the river meanders or zig-zags slowly across the plain. The river has lost much of its velocity and so much of its ability to carry heavier sand and silt in suspension. It therefore lays down part of its silt load as shingle beaches or sandbanks and builds up large flat plains by spreading alluvium over a wide flood plain or delta.

The second method divides the river up into zones based upon the presence or absence of certain fish species (see Frontispiece). These zones are:

The Headstream or Highland Brook which is formed from a number of small streams leading from a spring, a marsh or a glacier. The stream is small, shallow, has an irregular course, is often torrential with no pools and has a low water temperature. The only forms of plant life are mosses and liverworts. There are no fish in this zone.

The Troutbeck which is larger and more constant than the Headstream. The greater volume of torrential water in this zone cuts channels into the exposed rock floor or bedrock. The current is usually more rapid than in the Headstream and so the water, being also deeper, is capable of carrying objects in suspension. A typical troutbeck has a steep gradient (Fig. 3.1) and its sides are strewn with rough boulders and coarse pebbles. In sheltered parts of its course, where the flow is less, grit may be deposited. Because of the strong current and rocky conditions there is little plant growth. The water in this zone is always cold and oxygen-saturated. Areas of fast water alternate with irregular pools. The fish present in this zone are the brown trout (*Salmo trutta*), which is a powerful swimmer, the miller's thumb (*Cottus gobio*), and the stone loach (*Nemacheilus barbatula*) which shelter among the stones.

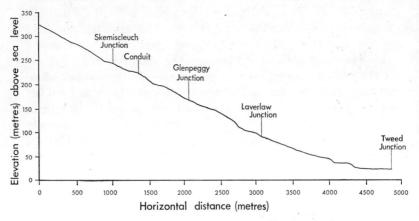

Fig. 3.1. Stream profile of Kirk Burn, Peebles-shire. (Vertical exaggeration × 10).

The Minnow Reach or Grayling zone has a less steep gradient than the Troutbeck. The current is therefore not so fast and, although the river still flows swiftly, the conditions are not torrential. There is therefore less erosion and some silt may be deposited in the quieter, smooth-flowing areas. In the areas with a moderate current filamentous algae may grow on the stones during the summer. Where silt is deposited other plants can gain a foothold and further silt accumulates among their roots. The presence of these quiet areas with water plants is characteristic of this zone. The water is still well-oxygenated but the water temperatures vary more. The areas of fast-flowing water or riffles now alternate more regularly with long pools. The fish which are characteristic of this stretch are the minnow (*Phoxinus phoxinus*) and, in some areas, the grayling (*Thymallus thymallus*). The fish of the Troutbeck zone are also present and so too is the eel and, in some parts of the country, the young of the salmon.

The last zone corresponds to the lower course of the river and is known as the *Lowland Course* or *Coarse Fish Reach*. Here the river is deep and slow-moving. The sluggish flow in this zone results in the deposition of silt, forming a muddy bottom on which many water plants can grow. The water temperature is more variable and the oxygen concentration less than that in the other zones. Although some of the fish characteristic of the upper reaches of the river may be present in this reach, conditions are not suitable for the successful completion of their life cycle. For example, the trout and salmon require silt-free gravel in which to lay their eggs, and cold, well-oxygenated water for their young. Conditions are now more suitable for other species of fish such as the cyprinids, the roach (*Rutilus rutilus*), the chub (*Leusicus cephalus* and the bream (*Abramis brama*). These fish can tolerate lower concentrations of dissolved oxygen and higher water temperatures and need water plants on which they can lay their eggs. In the very lowest reaches of a river running directly into an estuary or the sea, the flounder (*Platichthys flesus*) will occur during the summer months.

Not all rivers have these courses. The first zone obviously depends on the existence of hills which are fairly high and not permeable to water, as are chalk hills. Or the later ones may be missing and a torrential zone may run into the sea with no slackening of current. Rivers change during the passage of time and the changes are in the direction of shorter and less torrential mountain courses and longer and more meandering lowland courses. Table 4 shows the distribution of fish in the Tweed watershed. A distinct longitudinal succession occurs along a gradient of current velocity, highest in the upper reaches of the watershed, pH and alkalinity.

TABLE 4

The longitudinal distribution of fish in the Tweed watershed.

Station	1	2	3	4	5	6	7
Distance from sea (km)	90	87	84	69	58	32	26
pH	7·14	7·43	7·50	8·00	8·22	8·83	8·82
Alkalinity	28	47	62	75	99	168	250
Trout	x	x	x	x	x	x	x
Salmon		x		x	x	x	x
Eel		x		x	x	x	x
Stone loach				x	x	x	x
Grayling						x	x
Minnow						x	x
Roach						x	x
Gudgeon						x	x
Perch							x
Flounder							x

The physical nature of streams and rivers may be affected by the activities of man. Agricultural and forestry practices may have a profound effect on streams and rivers. One of the main agricultural practices which has an adverse effect on streams is land drainage. Land drainage causes a quicker run-off of water with an associated increase in bank erosion. This results in the transportation of a heavy silt load from the upper areas of streams and may also bring high accumulations of nutrients from agricultural fertilizers spread on the fields. Silt also occurs in rivers from erosion from overgrazing, muir burn and crop production on steeply sloping land. In Scotland sheep cause a great deal of erosion by destroying the ground vegetation when grazing. They also contribute to the instability of hill ground and river-banks by creating innumerable tracks and narrow paths and by using small knolls and irregularities in the ground and river-banks for protection from the weather.

After deforestation there is compaction of the soil and reduction of leaf litter followed by decreased permeability and storage capacity of the soil for water, and therefore a greater fraction of the water runs off the soil surface. This results in an increase in the absolute amount of water flow from that area, an increase in the flood peak discharge and a decrease in the low water flow. Increased run-off and flood peak flows usually mean increased erosion.

Suggested exercises

1. Make a tracing of a river in your area from an Ordnance Survey map. If possible, divide the river up into some or all of the zones mentioned.
2. Note the distribution of fish on your constructed map and note the presence of weeds, pools, etc.
3. From a contour map construct a profile diagram of the river.
4. Attempt to assess the fishing potential of the above river or stream. This can be done by sampling the habitat characteristics, as the fishing potential is strongly related to the condition of the fish habitat. Sampling is done by taking measurements along selected transects across a stream. The frequency of sampling should be every 200 metres. At every 200 metres a group of 5 transects is made, and the spacing between each transect is 2 metres. The results can give good estimates of stream length and width, surface area, pool and riffle area, depth and stream-bed composition, as well as of the stability and vegetative cover of the

TABLE 5

Stream sampling record as used by Herrington and Dunham

Drainage Unit No. _____

Sample No. _____

Field Crew _____

Date _____

Photo No. _____

Transect No.	Channel No.	Total width	Riffle width	Pool width	Width / Pool quality #1	#2	#3	#4	#5	Bottom material B	R	G	S-S	O	Vegetative bank cover F	B	O	Bank stability S	U	Stream channel depth A	B	C	Av.	Average gradient
					Metres										No. of banks					centimetres				Per cent
Total																								

TABLE 6

Pool quality recognition guide.

	Pool		
Quality class no	Length or width	Depth	Shelter[1]
1	Greater than a.c.w.[2]	60 cm or deeper	Abundant[3]
	Greater than a.c.w.	90 cm or deeper	Exposed[4]
2	Greater than a.c.w.	60 cm or deeper	Exposed
	Greater than a.c.w.	<60 cm	Intermediate[5]
	Greater than a.c.w.	<60 cm	Abundant
3	Equal to a.c.w.	<60 cm	Intermediate
	Equal to a.c.w.	<60 cm	Abundant
4	Equal to a.c.w.	Shallow[6]	Exposed
	Less than a.c.w.	Shallow	Abundant
	Less than a.c.w.	Shallow	Intermediate
	Less than a.c.w.	<60 cm	Intermediate
	Less than a.c.w.	60 cm or deeper	Abundant
5	Less than a.c.w.	Shallow	Exposed

[1] *Logs, stumps, boulders, and vegetation in or overhanging pool, or overhanging banks.*
[2] *Average channel width.*
[3] *More than $\frac{1}{2}$ perimeter of pool has cover.*
[4] *Less than $\frac{1}{4}$ of pool perimeter has cover.*
[5] *$\frac{1}{4}$ to $\frac{1}{2}$ perimeter of pool has cover.*
[6] *Approximately equal to average stream depth.*

(From: Herrington and Dunham, 1967.)

stream banks. The characteristics to measure, which are recorded on the form depicted in Table 5 are:

Width. The width of the water surface is measured to the nearest half-metre. Protruding rocks, stumps or logs are included as part of the total width. Note whether the channel is at high- or low-water level.

Depth. Depth is measured to the nearest centimetre at three points along each transect at intervals of one quarter, one half and three-quarters of the distance across each channel.

Pools. Five pool-quality classes are given in Table 6. These have been designed on the basis of pool size, water depth and fish shelter. The deeper and larger pools are considered

TABLE 7

Summary of stream drainage characteristics for the Kirk Burn, Peebles-shire.

		Total			Coniferous forest			Pasture and deciduous Woodland		
		No.	%	Area	No.	%	Area	No.	%	Area
Transects		40			21			19		
Adjusted stream lengths (km)		4·9			3·4			1·5		
Average width (metres)		2·3			2·2			2·3		
Surface area (hectares)				1·14			0·74			0·40
Riffle area			50·4	0·56		47·7	0·35		54·7	0·21
Pool area			49·6	0·58		52·3	0·39		45·3	0·19
Bank stability			60·0			45·2			76·3	
Proportion of pool area by quality	Class 1		0·0			0·0			0·0	
	Class 2		41·4			35·3			50·0	
	Class 3		0·0			0·0			0·0	
	Class 4		37·9			41·2			33·4	
	Class 5		20·7			23·5			16·6	
Proportion of bottom area by material	Boulder		22·7			24·0			19·6	
	Rubble		57·2			54·3			65·7	
	Gravel		7·9			9·5			3·4	
	Sand/silt		11·8			12·2			11·2	
	Other		0·0			0·0			0·0	
Proportion of stream bank vegetation type	Forest		50			80·9			15·8	
	Brush		12·5			7·1			18·4	
	Open		37·5			11·9			65·8	
Average depth (cm)		18·5			17·4			19·7		
Average gradient		4·7								

c

better fish habitat than the smaller, shallower and more exposed pools.

Bottom composition. Five types of bottom material defined as follows:

Boulder – Rocks over 300 mm in diameter.

Rubble – Rocks 50 to 300 mm in diameter.

Gravel – Rocks 3 to 50 mm in diameter.

Sand-silt – Particles less than 3 mm in diameter.

Other – Other matter (sunken logs or other debris).

Bank stability. Bank conditions at each end of a transect are rated either as 'stable' or as 'unstable'. An unstable rating is given if there is any evidence of soil sloughing within the past year. The number of stable banks for each transect is recorded as 0, 1 or 2. On multiple channels, only the two outermost banks are rated.

Streamside vegetation. Three types of streamside vegetation are recognized: 'forest', 'brush' and 'open'. Forest is defined as stands of trees. Other woody vegetation is defined as 'brush' and banks without woody types of vegetation are defined as 'open'.

Aquatic vegetation. At the time of recording the above features a note should also be made of the presence and abundance of aquatic plants.

An example of the results which will be obtained is given in Table 7 which summarizes the habitat characteristics of a small forest stream in Peebleshire.

Flows. These can be measured in a number of ways. The simplest is by multiplying the cross-sectional area by the speed of flow as measured by a floating object. Frequently an orange has been used as a float, as it is conspicuous and travels along almost entirely submerged. The following formula can also be used:

$$D = \frac{Wdal}{t}$$

when D = discharge, W = width, d = mean depth and l = distance over which the float travels in time, t. The term a is a coefficient which varies from 0·8 if the stream-bed is rough, to 0·9 if it is smooth (mud, sand, bedrock).

4 Plant Communities

Freshwater plants or water weeds can be divided into: (1) those that produce flowers, the *flowering* plants, and (2) those that do not produce flowers, the *flowerless* plants, such as the algae, mosses, liverworts and ferns.

Flowering plants

These are divided into three categories based upon where they grow:

(*a*) *submerged* weeds
(*b*) *floating-leaved* weeds
(*c*) *emergent* weeds.

(*a*) *Submerged weeds* The submerged weeds, along with the floating-leaved weeds, are the true water plants or *hydrophytes*. They are found only where there is water and soon die if exposed. They are commonly rooted in the mud, like the Canadian pondweed (*Elodea canadensis*) and Water milfoil (*Myriophyllum*), but a few are free-floating below the water surface, for example the Hornwort (*Ceratophyllum*). They are all completely submerged except when flowering, when most extend their flowering shoots above the water like the Mare's tail (*Hippuris vulgaris*), the Water violet (*Hottonia palustris*) and the Spiked water milfoil (*Myriophyllum spicatum*).

(*b*) *Floating-leaved weeds* Most of these are rooted in the lake- or river-bed and have long, pliable stems. A few, such as Duckweed (*Lemna*), Frogbit (*Hydrocharus morsus-ranae*), Bladderwort (*Utricularia*) and Water soldier (*Stratiotes aloides*), float freely at the water surface. The bladderwort is an interesting plant as it is carnivorous. The leaves are divided into hair-like segments on which are born bladder-like traps which catch and digest the prey.

The rooted plants in this category include the Amphibious bistort (*Polygonum amphibium*) and the Broad-leaved pondweed (*Potamogeton natans*). Many of the rooted plants may have submerged leaves as well as floating ones, like the Water crowfoot (*Ranunculus aquatilis*), the Starwort (*Callitriche stagnalis*) and Yellow water lily (*Nuphar lutea*).

(*c*) *Emergent weeds*   Emergent plants have erect aerial leaves arising from open water or mud. They grow in situations where the water level ranges from just below ground level to about half the maximum height of the plant. They are generally large and erect plants and most of the important ones have long, narrow leaves like grasses and are commonly called *reeds*. These plants form what is known as the *reed-swamp community*. This includes the Common reed (*Phragmites communis*), which often forms dense reed-beds around the margins of lakes and ponds, the Bur-reed (*Sparangium erectum*), the Reedmace or False bulrush (*Typha latifolia*), the True bulrush (*Schoenoplectus lacustris*) and the Yellow flag (*Iris pseuda-corus*). Other plants in this community are similar to broad-leaved plants and include the Water plantain (*Alisma plantago-aquatica*), the Arrowhead (*Sagittaria sagittifolia*), the Bog bean (*Menyanthes tri-foliata*) and the Marsh marigold (*Caltha palustris*).

Flowerless plants

These plants, too, can be divided into three categories, but not by where they grow so much as by their position in the plant kingdom. These categories are (*a*) the *algae* (*b*) the *mosses and liverworts* and (*c*) the *ferns* and *horsetails*.

(*a*) *Algae*   The two types of algae that interest us in freshwater are: (*i*) the Blue-green, which are the simplest forms of algae, and occur in gelatinous masses, in filaments, or as single cells. They may become so abundant in nutrient-rich waters in late summer that they form a 'scum' or 'water bloom'. Some of them produce poisons which are dangerous to fish: (*ii*) the Green algae, which is the largest group of freshwater algae. The main forms are the *desmids*, which are single-celled plants, some of which are free-floating and others attached to larger plants; and the *diatoms*, which, like the desmids, are single-celled but have a cell wall made of silica. Some are free-floating and others are attached to plants and stones. The diatoms also give rise to 'blooms' when conditions are suitable for their rapid growth. (*iii*) the Unbranched filamentous (*e.g. Spirogyra*) and the Branched filamen-tous (*e.g. Cladophora*) algae. These often form large floating masses or mats and are known as 'blanket weed' or 'cott'.

(b) *The mosses and liverworts*   There are several species of moss to be found covering the stones, particularly in fast-flowing streams. They are often the only plants present which provide shelter for many of the stream fauna. The Water moss (*Fontinalis antipyretica*) and Water feather moss (*Eurhynchium rusciforme*) are two of the better-known forms.

Liverworts are present on stones and boulders in and bordering streams. The two most commonly found are *Pellia* and *Marchantia*.

(c) *The ferns and horsetails*   There are only a few aquatic forms in this group of plants. Two of the commonest are the Quillwort (*Isoetes lacustris*) which grows submerged in the nutrient-poor lakes of northern Britain and the Water horsetail (*Equisetum fluviatile*) which is to be found emerging from the shallow water of bogs and ponds.

Let us look now at these plants in relation to their zonation. Plant succession is well demonstrated in ponds and lakes as there is a very distinct zonation in these larger aquatic plants with a succession that leads from open water to reed swamp and marshland. We can recognize four zones consisting of: (1) submerged plants, rooted and floating, (2) floating plants which are rooted, (3) emergent and erect plants, known as the reed-swamp plants, (4) erect plants, further from open water and known as marsh plants.

The first zonation occupies the deeper water beyond the zone of floating plants and extends down to depths of about 5 metres. The outer part of this zone may consist of plants which are not rooted in the mud but float freely, such as the duckweeds (*Lemna* spp.), frog-bit, hornwort and bladderwort. They occur where the depth of water is too great for rooted plants. The inner part of this zone includes the rooted plants which, as we have seen earlier, include the water milfoil and certain species of pondweed (e.g. *Potamogeton crispus*) and, in less eutrophic waters, the water violet and the stoneworts.

The second zone, closer to the shore than the first zone, consists of plants of the shallower water which are rooted in the mud, with some of their leaves floating or projecting above the water surface. Plants in this zone include the water crowfoot, the amphibious bistort, water lilies and some pondweeds (e.g. *Potamogeton natans*).

The third zone consists of the reed-swamp plants. On the outer fringe of the reed swamp is the bulrush (*Scirpus* or *Schoenoplectus lacustris*) which occurs in the deeper water. Nearer the shore is the common reed (*Phragmites communis*) which is the main swamp-

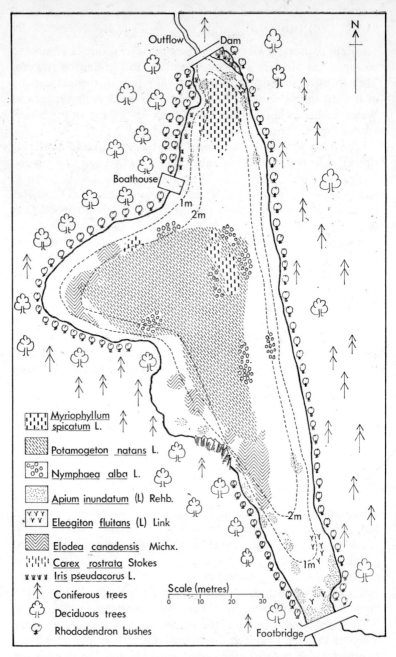

Fig. 4.1. Loch Dunmore, Perthshire (from: Morgan and Waddell, 1961).

forming species. It lives in 1–1·25 metres of water and grows rapidly by means of creeping rhizomes which penetrate the mud. As its leaves decay at the end of each year and silt accumulates, the water on the outer fringe of the swamp becomes shallower so allowing the plant to spread outwards. Other plants occurring in this zone are the reed maces and the bur-reed (*Sparangium*) which is sometimes the dominant plant along the banks of rivers and canals.

The last zone consists of the marsh plants such as the sedges (*Carex*) and rushes (*Juncus*), buckbean, water forget-me-not, brooklime (*Veronica beccabunga*) and meadowsweet (*Filipendula ulmaria*).

As the pond or lake becomes shallower so the first zones disappear and the latter zones extend outwards into the shallow water to replace them until the pond may disappear and be replaced by a reed-swamp, followed by a marsh and then a scrubby woodland or carr consisting of alders or willows. Finally this is replaced by the climax vegetation of the area, such as an oakwood. This succession does not always reach the climax stage as many physical factors will determine whether the lake or pond becomes shallower or not.

As we have seen, rooted plants are confined to water a metre or so deep. Some species can grow in less light than others and can therefore live in deeper water than others. Lower water temperatures and low light intensity may retard the growth of the plants, diminishing productivity and producing plants of reduced stature, particularly of those growing near the greatest depth limit. There are three photic zones in a lake distinguished according to the intensity of light received.

(*a*) A *photic* zone in which there exists sufficient light for the proper development of the higher plants. This zone may extend to a depth of 12 metres, as at Loch Fiart on the island of Lismore where the water moss *Fontinalis antipyretica* grows in water of all depths down to 12 metres. Usually the photic zone extends down to 4 metres or less.

(*b*) A *dysphotic* zone in which a few plants may exist, usually such plants are members of the lower cryptogams or non-flowering plants.

(*c*) An *aphotic* zone in which no light-demanding organisms can exist.

The type of bottom is also important in determining which kinds of plants can grow. Some plants prefer stones or sand, some silt, while others require organic matter. Wave action, water temperature, nutrient content and pH are other important factors, and as Welch (1952) points out, all these factors are so inherently involved with differences in depth that it is often difficult to evaluate the effect of any one of them. For example, many plants such as the common reed, bur-reed and water plantain always grow more luxuriantly when the mud is black and smelling of hydrogen sulphide; but plants such as the quillwort (*Isoetes lacustris*) and water lobelia (*Lobelia dortmanna*) cannot tolerate this type of bottom, not directly because of the presence of this type of mud but because of other factors such as the difference in the amount of nutrients. Other plants such as the bog-bean and sedge (*Carex rostrata*), however, are tolerant of a wide range of conditions. Again, in many districts, mountain lakes may be distinguished by the presence of plants such as quillwort and water lobelia, and the absence of reeds at the margin. In other areas this plant association may occur in lakes at a much lower elevation, while plants associated with richer lakes of lowland areas may be found in mountain waters. The reason for the presence or absence of certain plants is therefore not altogether one of elevation but is rather due to the supply of nutrients and the amount of exposure to winds, coupled with the nature of the shore.

Plant successions vary depending upon the type of silt and the rate of its deposition and Macan and Worthington (1951) give a number of examples:

(*a*) A coarse bottom, where the substrate is accumulating rapidly; for example, a bed of sand near the mouth of a stream is colonized first by *Juncus bulbosus* forma *fluitans*. As the substrate is built up, and the depth of water decreases, the floating-leaved *Potamogeton natans* gains a foothold. It in turn gives way to the emergent plants – the horsetail (*Equisetum fluviatile*) and sedge (*Carex rostrata*).

(*b*) A coarse substrate where there is slow accumulation of in-organic sediments will be colonized by the quillwort. In lakes such as Ennerdale this plant is common, and the only other abundant species is the stonewort (*Nitella opaca*). When the organic content of the soil is increased as vegetable debris accumulates and does not decompose, the quillwort is gradually replaced by *Juncus bulbosus* forma *fluitans*, water milfoil and water lobelia in shallow water. The floating-leaved and emergent species are the same as in the preceding succession.

(*c*) Where there is a fairly rapid accumulation of fine silt the succession is *Nitella* → *Potamogeton perfoliatus* → *Elodea* (the Canadian pondweed), *Sparangium minimum* (the small bur-reed), and *Potamogeton alpinus* → water lilies → *Phragmites*. This situation is typical of Windermere.

(*d*) In Esthwaite Water the rate of accumulation of fine silt is more rapid than in any other lake, and associated with this is *Najas flexilis* which is replaced by *Potamogeton pusillus* and other species of the same genus, having in common a long and narrow leaf. Water lilies are the plants of the floating-leaved stage, but the emergent typical of richly silted conditions is *Typha*, the reed-mace.

As Macan and Worthington point out, these four successions take place when the three environmental factors remain constant. They do not always remain constant; a rich supply of silt may be cut off or diverted so that a rich silted mud becomes organic, or the reverse may happen and an area where vegetable debris is accumulating and forming an organic substrate may come under the influence of silting which will bring about decomposition of plant remains. Tansley (1939) gives an interesting detailed account of these plant successions.

In rivers the type of plants present depends upon the rate of water flow and the presence of silt, the chemical content and the amount of light. In slow-moving waters where silt is present plants become established and gradually trap silt around the lower parts of their stems thus allowing them to spread. A distinct plant community may then become established. This may consist of a number of species of pondweed, water milfoil and yellow water lily. In fairly fast-flowing rivers plants with strong stems and deep roots stand up to the strain of the current. The water milfoil (*Myriophyllum alterniflorum*) and water crowfoot (*Ranunculus aquatilis*) can withstand very high water velocity.

Water plants can become a weed problem at times; they may impede navigation, hinder sport fishing and prevent commercial fishing. Weed control can be carried out in a number of ways. (1) Mechanical, by means of water scythes and weed cutters. (2) Biological, by the introduction of weed-eating mammals such as the coypu, or weed-eating fish such as the grass carp. (3) Chemical, with the use of copper sulphate for algae and herbicides such as dalapon and diquat for the rooted plants. There are, however, dangers inherent in the introduction of foreign species into a new environment as they may become pests, and be difficult to control. The coypu introduced into Norfolk caused serious damage to the river banks and had to be drastically controlled.

Suggested exercises

1. Plot (*a*) the distribution of plants in and around the pond or lake you have chosen to study. Note associations or communities of plants, the water depth in which they occur, the type of bottom and their time of flowering (see Fig. 4.1). To investigate the associations, set out the data you collect in a contingency table, and calculate χ^2 values. Determine the light penetration. For rough measurements of light penetration a secchi disc can be used. This is a white disc, 30 cm in diameter, which is lowered into the water to the depth at which it just disappears from sight, and then raised to the depth at which it just comes into sight. The extinction coefficient can be roughly determined by the following formula

$$K = 1.7/d$$

where *d* is the maximum depth in metres at which the disc is visible (Fig. 4.2).

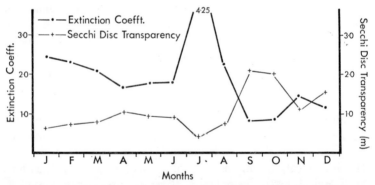

Fig. 4.2. Monthly extinction coefficient values and Secchi disc transparency readings on Duddingston Loch, Midlothian, 1970, (from: Mitchell, 1971). Each disc reading was made at 12.00 hours.

Note the ways in which the distribution of plants is related to the deposition of silt by inflowing streams, exposure to wave action, type of substrate and shade from overhanging trees. A note of the type of surrounding land use will also be most useful, for example grazing on sloping ground may cause soil erosion and result in more silt in the lake; heavy liming of arable land might result in an increased alkalinity, while the application of nitro-genous and phosphatic fertilizers will increase the nitrate and phosphate content of the water.

(*b*) Plot the distribution of plants (including algae) in a stream, river or canal. This is best carried out by means of a series of transects across the water and including the banks. Note the type of substrate, water velocity and light intensity.

2. Examine some bladderwort under the microscope and try to identify the organisms the plant has eaten. Does it feed at all times of the day and night?

5 Animal Communities

A chapter with such a title might justifiably imply that the reader can expect to see long lists of species that make up the different types of communities occurring in fresh waters. Suffice to say that there are many communities and that the species present vary according to the prevailing physical, chemical and biological factors. This holds whether the communities are planktonic (drifting in the surface waters) or benthic (living on the bottom).

This chapter will deal simply with the general types of animals living in lotic and lentic environments and the factors that influence their distribution. The reader is referred to the excellent guides by Macan (1959) and Engelhardt (1964) for the identification of fresh-water organisms.

Running waters

Eroding substrata In rapid streams with stony beds most animals are bottom-living. These types of streams often produce the largest variety and highest density of bottom-living organisms. Animals are adapted in a variety of ways to living in fast water. The limpet (*Ancylus*) and the leeches and flatworms attach themselves firmly to stones by their sticky undersurfaces; the caseless caddis larva of *Hydropsyche* and the larva of the black fly (*Simulium*) are anchored on silken webs attached to stones. In addition the black fly larva has a sucker at its posterior end. Many nymphs of stoneflies and mayflies have prehensile or grasping claws which enable them to cling to the rough stones or gravel. Some mayfly nymphs such as *Ecdyonurus* and *Rithrogena* have very flattened bodies which are held close to the surface of the large smooth stones on which they live, while other mayfly nymphs such as *Baetis* have streamlined bodies. Animals which rely entirely on their swimming ability, such as the freshwater shrimp (*Gammarus*) live in the crevices between stones and show a number of behavioural responses to current. Many stream animals show these responses in the way they orientate themselves to face into

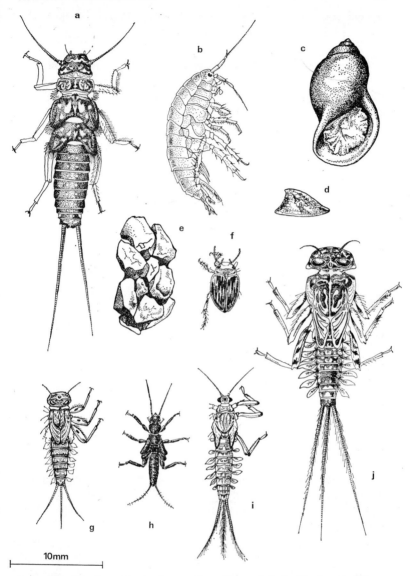

Fig. 5.1. Invertebrate fauna of eroding substrata (Scale: × 2½):
(*a*) stonefly – *Perla bipunctata* (*b*) freshwater shrimp – *Gammarus pulex*
(*c*) wandering snail – *Limnaea pereger* (*d*) limpet – *Ancylastrum fluviatile*
(*e*) caddis fly larva in case – *Agapetus spp.* (*f*) water beetle – *Oreodytes
rivalis* (*g*) mayfly – *Rithrogena spp.* (*h*) stonefly – *Nemoura spp.* (*i*) mayfly –
Baetis spp. (*j*) mayfly – *Ecdyonurus spp.*

the current so that they can use their gills and feed more effectively. Fig. 5.1 illustrates the main types of organisms found on eroding substrata of running waters.

Depositing substrata   The depositing substrata consisting of sand or soft silt are generally the least favourable for animals and support the smallest number of species and individuals. On the surface of the mud or silt the water louse (*Asellus*), the larva of the alder fly (*Sialis*) and various snails (*Valvata* and *Bithynia*) will occur, while in the silt the pea mussel will be found. In mud which is rich in organic matter only those organisms which can withstand low oxygen concentrations (e.g. *Tubifex* and *Chironomus*) exist, and in silt free of organic matter some of the silt-loving mayfly nymphs (*Ephemera* and *Caenis*) live. Where there are weed beds or where the current is very sluggish many more animals can find a niche in which to live, and dragon-fly nymphs, water boatmen and beetles and some mayfly nymphs (e.g. *Leptophlebia*) are common inhabitants. The main types of invertebrates to be found on depositing substrata are shown in Fig. 5.2.

Standing waters

The animals on a stony lake shore are on a substratum similar to the eroding substrata of running water, as they must be able to cling to or seek refuge beneath the stones when there is strong wave action. However, the genera and species of animals in such a lake environment may not always be the same as those in running water, for example, one species of mayfly of the genus *Ecdyonurus* occurs in lakes while two other species are confined to streams and rivers. While the genus *Ecdyonurus* is to be found in both lakes and streams the closely-related genus *Rithrogena* is confined to streams because it requires a flow of water to bring the correct amount of oxygen into contact with its body surface. For similar reasons other groups of animals may also be absent from lakes. *Simulium* and most of the net-spinning Trichoptera (e.g. *Hydropsyche*) are not found on lake shores as they depend on a water current to carry food to them.

The animals of sandy, silty or muddy shores are similar or identical to those occurring on the depositing substrata of slow-moving streams and rivers. Water beetles, water boatmen, chironomids (non-biting midges) and snails are the dominant organisms.

Although planktonic animals may be found in stagnant areas of rivers and streams they are generally confined to standing waters. The zooplankton consists mainly of copepods and Cladocera (water fleas,

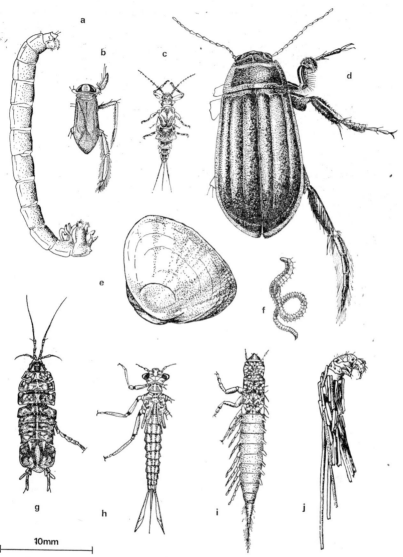

10mm

Fig. 5.2. Invertebrate fauna of depositing substrata (Scale: × 2½):
(*a*) chironomid larva of the large 'bloodworm' – *Chironomous thummi*
(*b*) water boatman – *Corixa dorsalis* (*c*) mayfly – *Caenis spp.* (*d*) great
diving beetle – *Dytiscus marginalis* (*e*) orb shell – *Sphaerium spp.* (*f*) red
worm – *Tubifex tubifex* (*g*) water louse – *Asellus aquaticus* (*h*) dragonfly
nymph – *Ishura spp.* (*i*) alder fly – *Sialis spp.* (*j*) caddis fly larva in case –
Limnephilus spp.

e.g. *Daphnia*). The species in the littoral area are different from those in the limnetic or offshore zone, in that the community in the latter is made up of few species but usually large numbers of individuals, and consists of copepods such as *Cyclops* and *Diaptomus* and highly transparent floating cladocerans such as *Sida, Bosmina, Polyphemus* and certain species of *Daphnia*.

Factors affecting distribution

The chief environmental factors affecting the distribution of aquatic animals in lakes and streams are:

The chemical nature of the water   This may affect the distribution of aquatic organisms in a number of ways. The concentration of dissolved oxygen is important and stonefly nymphs are only found in well-aerated water, while leeches, water lice and tubificid worms can live in waters of low dissolved oxygen content. The dissolved oxygen requirements of fish differ greatly from one species to another. Certain species, such as salmon and trout, require high dissolved oxygen concentrations, while other kinds, such as many fish from tropical regions, require only low concentrations.

The calcium content of the water is also important and the freshwater shrimp and many snails and mussels which are abundant in hard waters are less frequent in soft waters. Other groups of animals are also influenced by calcium content and in a study of the distribution of five species of triclads or flatworms in 122 lakes with a wide range of calcium concentration there was found to be a change in the percentage occurrence of the five species with various concentrations of calcium (Fig. 5.3). It has also been shown that the change in the mass of animals on plant detritus increases with the increase in the calcium concentration of streams (Fig. 5.4).

The physical nature of the water   Physical factors influencing animal distribution in freshwater include water movement and temperature. The surface velocity of water has been shown to have an effect on caddis larvae and it was found that the numbers of Trichoptera larvae along a current gradient became progressively less numerous with increasing distance from a mode (Fig. 5.5).

Water temperature is a critical factor in the life of fish and other aquatic organisms. It affects respiration, growth and reproduction. Each species of fish has a thermal tolerance zone in which it behaves in a normal manner. There is also a zone of higher temperature and

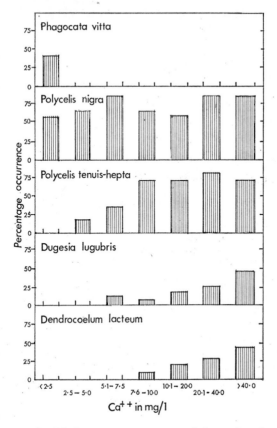

Fig. 5.3. Percentage occurrence of six species of triclad in waters with various concentrations of calcium (Reynoldson 1958) (Verh, int. Ver. Limnol. 13, p. 324, fig. 3) (from: Macan, 1963).

one of lower temperature, in which the species can survive for a certain length of time and above and below which the temperatures are lethal.

Fish can be divided into three groups according to their upper lethal temperature:

(a) fish with upper lethal limits below 28 °C, e.g. brown trout and char. These are called *stenotherms*, that is they can only live in water with a fairly narrow range of temperature,

D

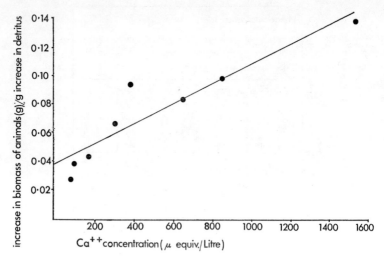

Fig. 5.4. Graph showing the relationship between the increase in biomass of animals/g increase in plant detritus and the calcium concentration of streams, with the calculated regression line $y = 0.0382 + 0.0000688x$ (from: Egglishaw, 1968).

(b) fish with upper lethal limits between 28° and 34 °C, e.g. salmon, pike, perch and roach,

(c) fish with upper lethal limits above 34 °C e.g. carp, tench and bream. These are *eurytherms*, that is they can tolerate a wide range of water temperature.

As regards the effect of water temperature on spawning, freshwater fish can be divided into three groups according to date of spawning:

(1) autumn and winter, when the water temperature is falling (below 10 °C).

(2) spring, when the water temperature is rising (10°–15° C), and

(3) summer, when the water is warm (above 16 °C).

Varley (1967) sets out a table (Table 8) to show that the fish species in each category have certain features in common and differ from those of the other categories, although there is some overlap. Important features are the size of the eggs, where they are laid, the time taken for them to develop, and the size and early habits of the fry.

TABLE 8
Summary of some features connected with spawning in
British freshwater fishes.

Spawning season	Autumn or winter (November/ February)	Spring (March/June)	Summer (May/July)
Daylength	short, increasing *or* decreasing	increasing	long, increasing *or* decreasing
Temperature	cold (up to 10° C)	rising (10–15° C)	warm (above 16° C)
Condition of rooted plants	died down	growing	full-grown
Location of eggs:			
laid among stones	whitefishes		
in nest in gravel	salmon, trout, char	grayling	
stuck to stones		minnow, barbel, chub, loach, gudgeon, bullhead	
among weeds		perch	ruffe
stuck to weeds		pike	roach, rudd, carp bream, tench
in nest of weeds			stickleback
Diameter of eggs	2·4 to 7 mm	1·3 to 3·2 mm	0·5 to 1·7 mm
Length of fry when read to feed	15 to 24 mm	4 to 8 mm	2·5 mm and more
Time between laying and:			
hatching of eggs	up to 5 months	4 to 28 days	3 to 20 days
fry ready to feed	up to 6 months	3 to 7 weeks	1 to 4 seeks
Food of young fry	invertebrates (same as older fish)	diatoms, water fleas and rotifers (different from food of older fish)	

(From: Varley, 1967)

The size, nature and stability of the stream bed As has already been mentioned at the start of this chapter, where the flow is rapid the bottom is composed of rocks, boulders or gravel, while where it is sluggish the bottom consists of mud and silt. The current speeds at which objects of a given speed are moved are given in Table 9. However, this classification is very simplified as much depends on whether the fall in the stream is regular or irregular.

The presence of vegetation The presence of vegetation in a stream provides an additional substrate on which animals can live and also a habitat for certain animals which might otherwise be unable to colonize the area because of the velocity of the water curent. In lakes,

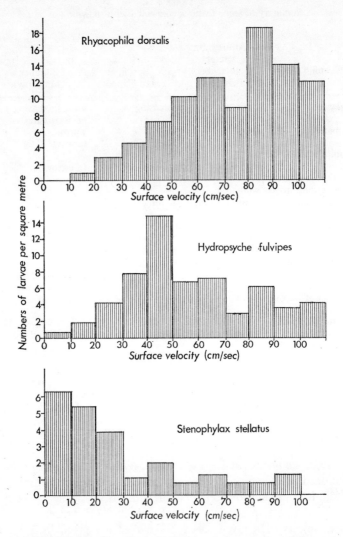

Fig. 5.5. Number of specimens per square metre at different current speeds (Scott 1958) (Arch. Hydrobiol. 54, pp. 350–351. Abb. 3, 6, 7) (from: Macan, 1964).

TABLE 9
Size of objects moved by different current speeds (Nielsen 1950b).

Speed of current	Diameter of objects moved	Classification of objects
10 cm/sec	0·2 mm	mud
25	1·3	sand
50	5	gravel
75	11	coarse gravel
100	20	pebbles
150	45	small stones
200	80	stones (fist size)
300	180	small boulders (man's head size)

(From: Macan 1963.)

vegetation is needed as a substrate upon which certain fish can lay their eggs.

Plant detritus, not only from aquatic vegetation but from terrestrial sources as well, is also important. Much of the plant detritus in a stream comes from sources outside the stream environment and is referred to as *allochthonous* plant material. This material provides a feeding substrate for micro-organisms and food for invertebrates. During its breakdown it also affects the chemical composition of the water. Egglishaw (1968) found that beech leaves collected in September and confined in nylon bags and placed in a stream lost 12·7 per cent of their dry weight in twenty-four hours. He considered that such quick losses are presumably due to leaching of soluble organic and inorganic materials, which probably provide a rich source of nutrients which can be assimilated directly by bacteria and some plants and animals.

Sampling methods

The simplest method of sampling the bottom fauna is with a fine-meshed net. In fast streams the stones are lifted in front of the net and the dislodged organisms will be swept into the net by the current. In lakes it is necessary to move the net through the water close to the bottom. If it is shallow enough to wade, the stones or weeds can be kicked to disturb the animals which can then be swept up with the net.

If we want to obtain quantitative samples so that we can compare the numbers and weight of animals present in each month or in different waters more elaborate sampling methods must be used. One of the simplest methods is collecting for a given time. Collecting is carried out for five or ten minutes during which the collector works slowly up and across the stream lifting stones and holding the net in such a way that anything beneath each stone is swept into it. Animals clinging to a stone are dislodged by vigorous washing in the mouth of the net and then the stone is discarded. While this is not very accurate it does produce useful figures and if the whole class is carrying out the sampling, the mean for the figures obtained can be used. One person should also examine stones for snails and limpets which will not be easily washed off. Another simple quantitative method is the 'kick' method, using a hand net consisting of a pyramid-shaped net, 30 cm deep, made of grit gauze, 12 meshes per cm, which is attached to a square frame with 24 cm sides. The net is held against the down-stream side of the area to be sampled and the bottom kicked with the collector's boot. Four rather prolonged kicks are made in an up-stream direction for a distance of about half a metre, each kick digging deeper into the bottom. Three of these 'four-kick' samples are taken, one in a pool, one in a riffle and a third in a position where conditions are intermediate between those at the other two sampling sites.

A more accurate quantitative method is that using the shovel sampler which is pushed forward through the substrate. The net is rather like a shovel open at the back and carrying a bag. In one model the net frame is 30 cm wide and 45 cm high and the lower part of the frame, which is in contact with the substrate, has a cutting edge. Strong, small-meshed netting is attached to the frame, to prevent the inner net from being worn and torn by the stony substrate. Most of the organisms collect in the inner net which is made of 60 meshes per 25 mm grit gauze. Over this net is attached an inner metal frame with a large mesh net of thick twine. This prevents the larger stones falling into the gauze net. The shovel sampler is pushed forward into the substrate a given distance, to sample a given area of bottom. Any stones collected on the inner frame of large-meshed netting are lifted out and washed in a basin of water. An addition to this sampling technique is a wooden three-sided sampling frame with raised sides. The two arms are of solid wood while the upstream side is made of weld-mesh. The purpose of the frame is to outline the sampling area and to prevent gravel and small stones, together with some of the fauna, being pushed ahead of the net. The shovel net is of limited use in streams where a considerable proportion of the bottom is composed of large stones or bedrock.

A number of people studying bottom fauna have found that the distribution of aquatic insects in a stream is not random, as they may be found in concentrations where detritus (Fig. 5.4) has accumulated under stones or in eddies. Furthermore, considerable variation has been found between adjacent samples and seasonal variations in numbers are also considerable.

Quantitative bottom samples can be collected from lakes by means of various types of spring-loaded grabs which close, either on contact with the bottom or with the aid of a messenger which triggers off a closing device, and scoop up a known area of substrate. Core sampling is also used to sample lakes with a muddy substrate.

Plankton can be sampled qualitatively using a net made of bolting silk, which has a very fine mesh. For zooplankton the silk should have 60 meshes per 25 mm, and for phytoplankton 180 meshes per 25 mm. For quantitative work a large conical net is used which is towed horizontally, obliquely or vertically through the water. A flowmeter is attached to the mouth of the net to measure the amount of water passing through the net.

Sampling of fish populations is described in Chapter 8 pages 65 to 73.

Suggested exercises

1. Collect qualitative samples of organisms from (*a*) a pond (*b*) a stony lake-shore (*c*) a muddy lake shore (*d*) a fast-flowing stream (*e*) a slow-moving stream (*f*) weed beds. Make a note of the animals in the various communities and draw any which show special adaptations (for feeding, respiration, reproduction and attachment) to the habitat in which they live.

2. Take monthly quantitative samples of organisms from a stream and note the differences in the abundance of various groups of animals. You will already be recording the physical and chemical characteristics of your stream and these will help you to correlate your results and compare them with data for other waters. A recent publication which will help you to choose the correct statistical method for analysing your results is *Some Methods for the Statistical Analysis of Samples of Benthic Invertebrates* by

J. M. Elliott, 1971. (Freshwater Biological Association, Scientific Publication No. 25.)
3. Determine the abundance and species composition of the zooplankton in a pond or lake over the year.

The Freshwater Biological Association at Ferry House, Ambleside, Westmorland, sell very good sampling nets which can be used for some of the qualitative and quantitative work described. The Association also has a wide range of Keys to the various groups of freshwater animals. They are very reasonably priced and would be a useful addition to the laboratory bookshelves.

6 The Flow of Energy

Now that we have looked at the types of freshwater environment, the plants and animals inhabiting them and the factors controlling their distribution, it is appropriate to draw our knowledge together and consider the ways in which the energy flows through this ecosystem.

The ecosystem

The ecosystem is made up of four constitutents. (*a*) The *abiotic* substances, which are the basic elements and compounds of the environment, such as water, carbon dioxide, oxygen, calcium, nitrogen and phosphorus salts and humic acids. (*b*) The *producers*, mostly green plants, known as autotrophic organisms because they are 'self-nourishing', and are able to fix light energy and manufacture food from simple inorganic substances. The greatest autotrophic metabolism occurs in the upper layers of the water where light energy is available. (*c*) The *consumers* or macro-consumers, being mostly animals which ingest plants or other animals or particulate organic matter. The consumers are *heterotrophic* ('other nourishing') organisms which use and decompose the complex materials synthesized by the producers or autotrophs. Most heterotrophic activity takes place where the organic matter accumulates in the soils and sediments. (*d*) The *decomposers* or micro-consumers, sometimes known as saprophytes, which are heterotrophic and consist chiefly of the fungi and bacteria that break down the complex compounds of dead protoplasm, absorb some of the decomposition products and release simple substances which can be used by the producers (Fig. 6.1).

The food chain

Odum has defined a food chain as 'The transfer of food energy from the source in plants through a series of organisms with repeated eating and being eaten' (Odum, 1959). Food chains lead usually from smaller to larger forms and start from herbivorous and scavenging

49

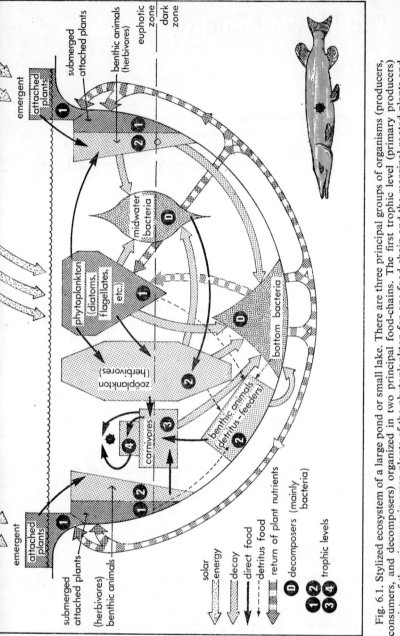

Fig. 6.1. Stylized ecosystem of a large pond or small lake. There are three principal groups of organisms (producers, consumers, and decomposers) organized in two principal food-chains. The first trophic level (primary producers) consists of the microscopic green plants of the phytoplankton for one food-chain and the marginal rooted plants and their epiphytes for the other. The corresponding second-trophic-level animals are filter-feeding zoöplankton and browsing benthic herbivores, respectively. The third and fourth trophic levels of carnivorous animals could lead to an old cannibalistic pike operating at both the fourth and fifth trophic levels (from: Russell-Hunter, 1970).

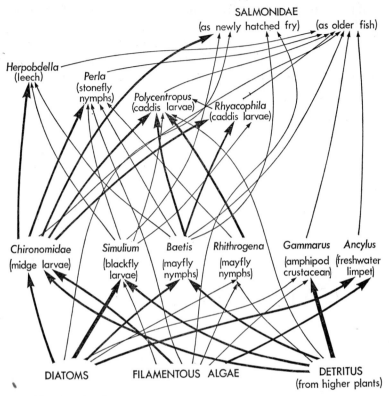

Fig. 6.2. The relatively complex food-web of trout and similar salmonid fishes in a stream community. Note that an important primary food resource is in the form of detritus derived from the productivity of land vegetation (that is, allochthonous to the stream itself). Increased complexity of links in such a food-web can probably create increased stability for such a stream community. (Based on data tabulated for streams in the Dee river system, Wales, by R. M. Badcock in J. Anim. Ecol., 18: 193–208, 1949) (from: Russell-Hunter, 1970).

forms which in turn depend directly or indirectly on plants e.g. Plant → Herbivore → Carnivore. Organisms whose food is obtained from plants by the same number of steps are said to belong to the same *trophic level*. Because organisms in any trophic level may obtain their food from more than one source, food chains can be very complex and are, for that reason, often referred to as food webs. The food web of a small stream community in Wales is shown in Figure 6.2. This shows the interlocking nature of food chains. We can add to this food web by introducing those organisms at the top of the food chain known as the 'top carnivores', such as man, the otter, the heron and the cormorant, which prey on the trout.

Ecological pyramids

There tends to be a decrease in the numbers of animals as one moves up the food chain, so that the animals at the bottom of the chain are relatively abundant while those at the top are relatively few in number. This decrease in numbers at each trophic level produces what is known as the *pyramid of numbers* or ecological pyramid. The pyramids differ in shape depending on whether the primary producers are large or small (Fig. 6.3). Ecological pyramids may be of three general types (*i*) the *pyramid of numbers*, in which the number of individual organisms (no./m²) is depicted, (*ii*) the *pyramid of biomass* based on the total dry weight (g/m²), or calorific value, (*iii*) the *pyramid of energy* in which the rate of energy flow (kcal/m²yr)

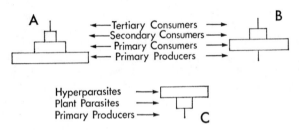

Fig. 6.3. Pyramid of numbers. A, when the primary producers are large, B, when the primary producers are small; C, a plant/parasite food web (from: Phillipson, 1966).

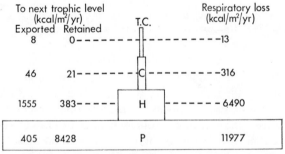

Fig. 6.4. Pyramid of energy for Silver Springs, Florida, U.S.A. (Based on Odum, 1957, courtesy of Ecol. Monogr.).

at different levels is shown. The pyramids of numbers and biomass are limited in that they reveal only the amount of material present at one moment of time and this amount is referred to as the *standing crop*. They give no indication of the total amount of material and the rate at which it is produced. For this reason, the pyramid of energy is more useful as it gives the amount of energy utilized by the organisms in the different trophic levels in a given area over a fixed period of time (Fig. 6.4).

Energy flow

Let us now look at the way energy flows through the aquatic ecosystem. Fig. 6.5 is a simplified energy flow diagram constructed by Odum (1959). The boxes represent the population mass or biomass and the pipes depict the flow of energy between the living units. Only about half of the average sunlight impinging upon green plants (the producers) is absorbed by photosynthesis, and only a small portion of absorbed energy (about 1 to 5 per cent) is converted into food energy. The total assimilation rate of producers in an ecosystem is called *primary production* or primary productivity (P_G or A in Fig. 6.5), and is the total amount of organic matter fixed, including that used up by plant respiration during the measurement period. Net primary productivity is the organic matter stored in plant tissues in excess of respiration during the period of measurement. Net production represents food potentially available to the consumers or heterotrophs. In Fig. 6.5 net primary production is represented by the flow P that leaves the producer component. At each transfer of energy from one organism to another, or from one trophic level to another, a large part of the energy is degraded into heat and it can be seen that the energy flows are therefore considerably reduced with each successive trophic level. Fig. 6.6 shows the pattern of energy flow in a fish. The energy taken in as food by a fish is disposed of in three ways: part of the energy is lost in faeces and urine in the process of egestion and excretion; a larger part is used up in the metabolic or day-to-day activities of the fish, the energy being liberated in the process of respiration (a formula is given in Fig. 6.6 for this component based on the weight of the fish and the average river temperature in each month of the year) and the rest of the energy ingested is stored in the body as new flesh, and is measured as growth. The total energy requirements of the fish are, therefore, growth, respired energy and energy of excretion. Based on the energy requirements of this fish a diagram has been constructed by Mann (1946) to show the pattern of energy flow in the bottom fauna and fish populations of the River Thames (Fig. 6.7).

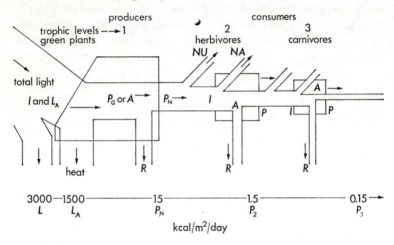

Fig. 6.5. A simplified energy flow diagram. The boxes represent the standing crop of organisms (1: producers or autotrophs; 2: primary consumers or herbivores; 3: secondary consumers or carnivores) and the pipes represent the flow of energy through the biotic community. L = total light; L_A = absorbed light; P_G = gross production; P_N = net production; I = energy intake; A = assimilated energy; NA = non-assimilated energy; NU = unused energy (stored or exported); R = respiratory energy loss. The chain of figures along the lower margin of the diagram indicates the order of magnitude expected at each successive transfer starting with 3 000 kcal of incident light per m² per day (from: Odum, 1963).

A schematic representation to show the main processes involved in fish production is given in Fig. 6.8. These are:

(*a*) The production of food for the fish being considered, including the population and production dynamics of the food organisms.

(*b*) The feeding of the fish and the assimilation of some of this food, the remainder being egested as faeces.

(*c*) The loss of some of the matter and energy assimilated in metabolic processes including excretion.

(*d*) The growth of the individual fish.

(*e*) The sum of the growth of all the individuals in the stock, or the production, including the production of gonad products.

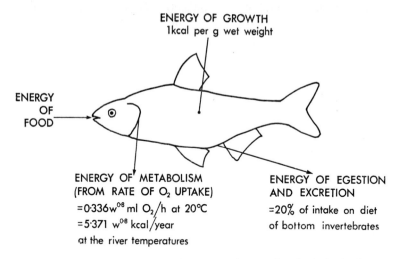

Fig. 6.6. A diagram showing the pattern of energy flow in the body of a fish (from: Mann, 1964a).

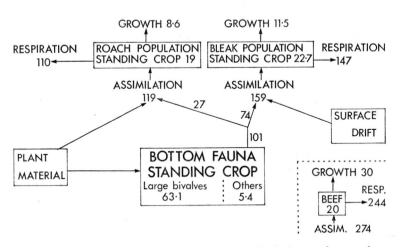

Fig. 6.7. Diagram of the pattern of energy flow in the bottom fauna and fish populations of the River Thames. Figures within boxes are the calorific values (kcal/m²), other figures kcal/m²/yr. In the bottom right-hand corner is shown for comparison a set of figures for beef raised on grassland (from: McFadyn) (from: Mann, 1964b).

(*f*) The accumulation of this production as stock or biomass which in turn suffers loss through mortality of several kinds. The principal sources of mortality are:

(*i*) Fishing, giving yield to man.

(*ii*) Predation, leading to production of predators.

(*iii*) Infection by parasites and pathogens leading to the production of parasites. (Some parasites remove material and energy from the system at other stages, e.g. between the ingestion and assimilation of food and growth. Many parasites may not cause the death of fish.)

(*iv*) Other causes of mortality, including diseases not due to parasites or pathogens, accidents and pollution.

(*g*) Some of the organic matter in all the fish that die will return as either organic matter or mineral nutrients to the lower trophic levels.

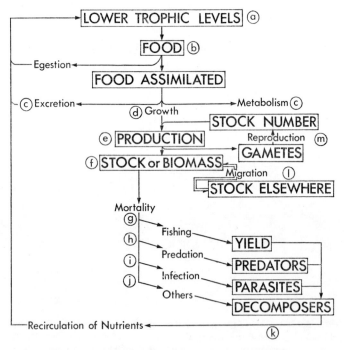

Fig. 6.8. Diagrammatic representation to show the main processes in fish production (from: Backiel and Le Cren, 1961).

(*h*) As well as natality and mortality, emigration and immigration will have similar negative and positive effects upon the stock.

(*i*) Reproduction or replenishment of the stock with new individuals, is an essential process. Production contributes to this in the form of the matter (and energy) contained in the fertilized egg.

We will consider methods for the measurement of production in Chapter 9.

Suggested exercises

1. Set up an aquatic ecosystem in an aquarium with water plants, snails, caddis larvae and other invertebrates and fish. Try not to introduce too many carnivores. Record the behaviour of the various animals.

2. Construct a food web in the pond, lake or stream you have under study. It will be necessary to examine the stomach contents of the fish and the gut contents of the invertebrates several times in the year to be certain that you have recorded most of the food items of the various organisms.

3. Try and build a pyramid of numbers and a pyramid of biomass using the various methods of sampling described in Chapters 5 and 8. Quantitative sampling of the producers may be difficult. The higher plants (i.e. the macrophytes) can be removed from known areas and weighed. Removal may be direct by hand from a quadrat. In a stream the filamentous algae can be scraped off the substrate and weighed. In a lake the diatoms and desmids can be collected from a known volume of water (say 10 to 20 ml) by centrifuging for 15 to 30 minutes at 1 500 rpm. The sample can then be weighed. To obtain an estimate of their numbers sub-sampling will be necessary – the diatoms in 1 ml of the liquid being counted on a gridded slide, or in a haemocytometer. A good method of estimating the number of algae is given in a paper entitled: The inverted microscope method of estimating algae numbers and the statistical basis of estimates by counting, by Lund, Kipling and Le Cren, in *Hydrobiologia*, XI, **2**, 143–170; 1958.

Setting up an aquarium   Setting up an aquarium is quite a simple process and is most enjoyable.

Aquaria come in a wide range of sizes and one's choice will be

E

governed by funds available. The best aquaria have a framework made of angle iron and can be obtained from aquarists or pet shops. The bottom of the tank should be covered with fine gravel to a depth of 3 to 6 cm. If fine gravel is hard to find it can be bought at aquarists. The bought gravel will need a good wash before being put in the aquarium tank. The tank can now be filled with water to within 6 to 10 cm of the top. Larger stones can then be added. Good stones to use are those from streams or lakes bearing the water moss *Fontinalis*. This is a very good plant for aquaria. Other plants can also be planted in the gravel. A little of the floating duckweed *Lemna* on the water surface is quite useful but as it spreads rapidly it will need controlling.

A source of oxygen is necessary, and this is best provided by a pump. The Hi-Flo piston pumps are very good and make no noise. The pump is plugged into a power source and plastic tubing is attached to the pump outlets. An air stone or diffuser is inserted into the other end of the tubing which is placed in the tank.

Fish, snails, caddis larvae and water beetles can now be added. Good aquarium fish are roach, minnows and perch. If sticklebacks are considered they should be used on their own as they can inflict nasty wounds on other fish, particularly during the breeding season. However, they are interesting fish for an aquarium and exhibit a number of display patterns. For example, if a small mirror is held in front of the male stickleback when in its breeding colours it will undergo threat display in front of its image. The fish will need to be fed on small worms, water fleas (*Daphnia*) or fish pellets (there are a number of proprietary brands). Most duck-ponds are a good source for *Daphnia*.

A disease sometimes encountered in aquarium fish is white-spot. The best method of preventing an attack of white-spot is to quarantine new stock for a few days. The average incubation period for the disease is four to five days so that fish which show no sign of the disease after a week can usually be considered disease-free. The best method of preventing white-spot, which is often found on minnows, is with Liquitox (obtainable from aquarists) or treating it with 1 part in 2 000 of formalin for five or ten seconds, or brush with weak paraffin.

If one wants to catch the caddis flies or mayflies when they emerge as adult insects it is advisable to place fine-mesh netting over the top of the tank.

During the summer months algal growth will cover the glass sides of the tank and the water may become a green 'soup'. The snails will help to remove the algae but it will be a 'losing battle' and there is nothing for it but to change the water fairly regularly and scrape the glass sides. There are algicides on the market but they may be slightly toxic to fish. Filters also help and can be purchased from an aquarist. The high-speed filters which have glass wool and two grades of charcoal are very good.

If there is any doubt about the toxicity of the water supply have it tested by the public analyst or local river authority. For example, if the water comes through new copper piping it is likely that the concentration of copper will be too high for fish. Copper piping that has been installed for some years is usually safe. High concentrations of chlorine are not good, but such concentrations are not likely in tap water. If the water supply is not suitable for fish then local pond or river water will have to be used.

7 The Biogeochemical Cycle

So far we have considered the orders of magnitude regarding energy flow in the aquatic environment. However, the movement of nutrient materials in the system is an equally important consideration. The movement of these materials is cyclical, and there are three major types of cycles. One of these is the water or *hydrological* cycle, which involves the movement of a compound, and the other two are chemical cycles. The chemical cycles involve biological organisms and their geological environment and are referred to as *biogeochemical* cycles. In one group of these the atmosphere constitutes the major reservoir of the element, and these cycles are known as gaseous nutrient cycles, of which carbon and nitrogen are the main representatives. In the other group the reservoir of the elements is made up of the sediments and the cycle is known as the sedimentary nutrient cycle. The essential elements of this group are sulphur and phosphorus.

The hydrological cycle

We can consider all water as being involved in a cyclical movement. Starting with the atmosphere (Fig. 7.1), water vapour in the form of clouds condenses and may give rise to rain (precipitation). Not all the rain reaches the earth's surface as some evaporates while falling and some is caught by vegetation or the surfaces of buildings and evaporates. The rain which reaches the ground may (*i*) remain on the surface as pools and surface moisture which eventually evaporates, (*ii*) flow over the soil surface as surface run-off into streams and lakes, (some of this will either evaporate into the atmosphere or seep into the ground towards the groundwater, or eventually flow into the sea) (*iii*) either sink into the ground and be used by plants in the process of transpiration or percolate to the underlying groundwater where it may remain for very long periods of time, eventually being removed by seepage into the streams. Most surface drainage and groundwater flows into the sea and consequently the sea is the major contributor of water vapour to the atmosphere through evaporation, which is not surprising when one considers that the seas cover 71 per cent of the world's surface area.

In some areas of the world the hydrological cycle has been modified by the activities of man. For example, drainage schemes have reduced water table levels and increased streamflow; irrigation has increased soil moisture content and loss by evaporation; afforestation and deforestation has altered the surface run-off pattern, and dams and reservoirs have controlled and altered the pattern of natural river-flow.

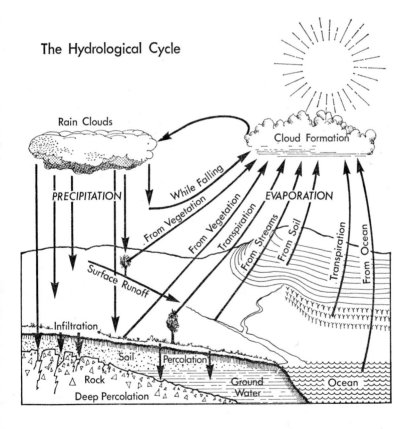

Fig. 7.1. The Hydrological Cycle (from: The Yearbook of Agriculture, Water, 1955. U.S. Dept. of Agriculture).

The biogeochemical cycle

In Fig. 7.2 a biogeochemical cycle is superimposed on an energy-flow diagram to show the interrelation of the two processes. The essential elements very rarely occur evenly throughout the environment; more often than not they occur in compartments or 'pools' with varying rates of exchange between them. It is useful to distinguish between a large, slow-moving non-biological 'pool' and a smaller but more active 'pool' that is exchanging rapidly with organisms. In Fig. 7.2 the large reservoir is the box labelled 'pool' and the rapidly cycling material is represented by the stippled circle cycling between heterotrophs and autotrophs.

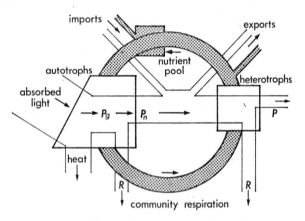

Fig. 7.2. A biogeochemical cycle (stippled circle) super-imposed upon a simplified energy-flow diagram, contrasting the cycling of material with the one-way flow of energy. P_g = gross production: P_n = net primary production, which may be consumed within the system by heterotrophs or stored or exported from the system (from: Odum, 1963).

For our example of a biogeochemical cycle in an aquatic eco-system we will take the sulphur cycle (Fig. 7.3). Sulphate in the water is taken up by plants and, in turn, by animals. When the bodies of plants and animals are decomposed hydrogen sulphide is produced. Some of this is then reconverted to sulphate by special sulphur bacteria. As much of the decomposition takes place in the sediments, anaerobic (i.e. without oxygen) conditions tend to develop, with the result that much of the hydrogen sulphide may not be oxidized but pass into the reservoir pool. The sedimentary part of this cycle involves the precipitation of sulphur in the presence of iron under these anaerobic conditions. Ferrous sulphide is insoluble in neutral or alkaline water and so sulphur is 'locked up' under these conditions for varying periods of time. However, when iron compounds are formed phosphorus is converted from the insoluble to the soluble form and so becomes available to organisms in the water. So we can now see that the reduction of sulphate and sulphide in the anaerobic mud helps to regulate the cycling of the nutrients.

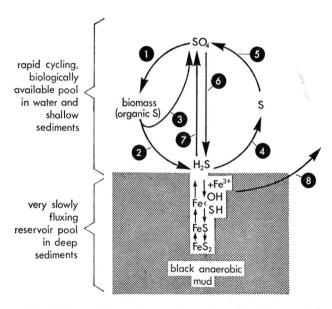

Fig. 7.3. The sulphur cycle as an illustration of the movement of a specific element in a specific ecosystem. Organisms play key roles in the rapidly cycling pool as follows: (1) primary production by autotrophs; (2) decomposition by heterotrophic microorganisms; (3) animal excretion; (4), (5) steps by specialized colourless, purple and green sulphur bacteria; (6) desulphovibrio bacteria (anaerobic sulphate reducers); (7) thiobacilli bacteria (aerobic sulphide oxidisers). Step 8 releases phosphorus (from insoluble ferric phosphate), thus speeding up the cycling of this vital element (from: Odum, 1963).

8 Populations

A great deal of information on life in freshwater can be obtained from a study of populations of invertebrate organisms or fish. Population studies are very rewarding and full of interest, as usually the study of one aspect of a population of animals leads on to another. If we are looking for a definition of a population we can use that given by Odum: 'All of the organisms of the same species found occupying a given space.'

There are certain characteristics of a population which should be noted before making any population study. These are:

Density – the population in relation to a unit of space (e.g. no./m^2, g/m^2 or kg/Ha).

Recruitment or natality – the rate at which new individuals are added to the population by reproduction.

Mortality – the rate at which individuals are lost by death.

Age composition – the proportion of different individuals of different ages in the group.

Population dispersion – the way in which individuals are distributed in space.

Population growth rate – the net result of recruitment, mortality and dispersal from the group.

Let us look at some of these characteristics in more detail.

Population density

The density of a population should always be expressed in terms of a definite amount of space, and preferably it should also be expressed in more than one term. For example, number per square metre tells us how many animals are present but little more, and for comparative purposes this is not sufficient. The additional information on the weight of those animals would be of very great value. If we look at Table 10, which records the density and weight of a number of species of fish in the same stretch of stream over a five-month period, we see how two figures per unit area reveal very much more about the fish populations than would one.

TABLE 10

Density (D) (No./m²) Standing crop (S) (g./m²) of fish in a 25 metre long stretch of stream (8 metres wide) on the Eden Water, 1969. Stream bed consists of medium to small-sized gravel with beds of water buttercup; stream bordered by bur-reed and alder.

	June		July		August		October		November	
	D	S	D	S	D	S	D	S	D	S
Trout	0·07	3·3	0·10	4·5	0·13	4·5	0·20	10·9	0·21	9·6
Salmon	0·05	0·5	0·05	0·8	0·09	1·4	0·22	6·2	0·12	2·4
Grayling	—	—	0·02	1·0	—	—	—	—	—	—
Roach	—	—	0·005	1·2	—	—	—	—	—	—
Minnow	—	—	—	—	—	—	0·005	+	0·005	+
Stone Loach	0·03	0·1	0·20	0·7	0·17	0·73	0·03	0·16	0·02	0·14
Gudgeon	—	—	—	—	—	—	0·005	+	—	—
Lamprey	—	—	0·01	+	0·005	+	—	—	—	—
Eel	0·50	10·2	0·50	9·3	0·37	9·2	0·47	15·1	0·01	3·7
All species	0·65	14·1	0·88	17·5	0·76	15·8	0·92	32·4	0·46	15·8

+ = small amount
— = not present

(from: Mills, 1970)

Natality and mortality

These are complex population characteristics which for most situations require a strict mathematical treatment beyond the scope of this book. For those wishing to study this aspect of population dynamics a very useful treatment of the subject is given in *Methods of Assessment of Fish Production in Fresh Waters* edited by W. E. Ricker and published in 1968 by Blackwell. However, mortality and survival can be estimated from the formula

$$S = n \sqrt{\frac{R(a+n)}{Ra}}$$

in which
S = the average level of annual survival between age a and age $a+n$
R = the number of fish of age a
$R(a+n)$ = the number of fish of $a+n$.

Assuming a fish population with the following distribution per age-group:

$0+ = 2\ 136;\ 1+ = 862;\ 2+ = 352;\ 3+ = 60;\ 4+ = 6$

Then of age $0+$ to age $4+$:

$$S = 4\sqrt{\frac{6}{2\ 136}}$$
$$= 4\sqrt{0 \cdot 002809}$$
$$\simeq 0 \cdot 23$$
$$\text{or } 23\%$$

The level of annual mortality: $Z = 1 - S = 77\%$

This of course assumes that the recruitment has varied little in the course of the preceding years, and that the migrations of fish have been of little importance.

Age composition

We are fortunate in our study of populations of freshwater animals to have such a useful animal as the fish which is a good subject for the study of age distribution. This is because fish continue to grow throughout their life-span and their age can be determined from growth rings on their scales (Fig. 8.1). The method most commonly used for reading scales is to wash a number of scales with a camel-hair paint brush, to remove the skin covering the scales, and then to mount them on a glass slide under a large coverslip. The slide is then placed on a microscope stage under the microscope lens and the magnified image projected on to a screen. The magnified images of the scales are then examined and the one showing the clearest scale pattern selected, first making sure that it is not a regenerated scale which does not give the complete age of the fish as its centre is filled with regenerated scale material. The positions of the annuli (or 'bands' of winter rings) are marked on a strip of cardboard laid on the image on the screen from the centre of the scale to its edge. The strip of cardboard is then laid on a piece of graph paper on which the expected range of fish lengths has been marked along the ordinate (Y-axis). With the zero of the scale measurement at the origin, the strip is rotated until the scale edge mark is opposite the observed length of the fish. Intermediate lengths corresponding to the annuli can then be read on the ordinate side of the graph paper. The data from such scale readings can then be incorporated into the construction of a growth curve such as that depicted in Fig. 8.2. This procedure is possible when there is a direct proportionality between the

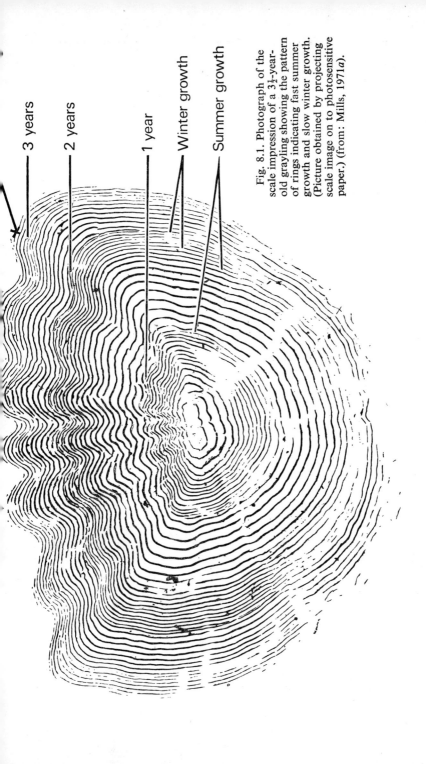

3 years

2 years

1 year

Winter growth

Summer growth

Fig. 8.1. Photograph of the scale impression of a 3½-year-old grayling showing the pattern of rings indicating fast summer growth and slow winter growth. (Picture obtained by projecting scale image on to photosensitive paper.) (from: Mills, 1971a).

rate of growth in length of the fish and that of the scales, as is the position in trout and many cyprinid fish.

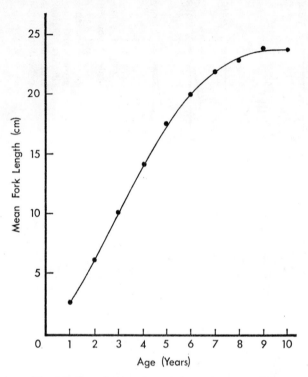

Fig. 8.2. Growth rate of roach in Loch Lomond (from: Mills, 1971b).

A much simpler method of age determination is the analysis of size frequency distributions. This method uses the individual lengths of a large number of fish of a population. It requires a unimodal size distribution of all fish of the same age, and is easy to use if there is no overlap in the size of the individuals in the adjacent age-groups or difference in the growth-rates of males and females. However, usually it can be applied only to the youngest age-groups of a population. In Figure 8.3 depicting the length frequency of roach the first three age-groups are quite distinct: the one year old fish showing a length range of 2·5 to 4·5 cm; the two year old fish a length range of 5·0 to 8·0 cm, and the three year olds from 8·5 to 11·5 or 12·0 cm. After 12·0 cm, the picture is difficult to interpret as, while there are peaks at 13·5 and 16·5 cm, there are at least five age-classes between these lengths. It may be possible in length frequency distributions to trace the fate of an age class over several months if not years and in Figure 8.4 it is possible to identify the earlier age-groups of pike. The one year old group can be traced through from April to August and identified in May of the following year in the two year old group. In 1959 the group is particularly noticeable while in 1960 it is not so well-marked and in 1961 is poorly represented. This is also apparent in the two year old group, but the older age-groups make it difficult to follow the fate of this particular age-group. The older age-groups are obscured by differences in growth-rates of males and females and by the overlapping which occurs in the length distribution of adjacent age-groups of the same sex.

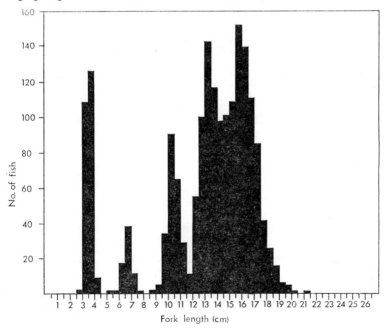

Fig. 8.3. Length frequency distribution of roach caught in Humbie reservoir (from: Mills, 1971*b*).

To determine the proportion of individuals of different ages in the population it is necessary to obtain estimates of the population size.

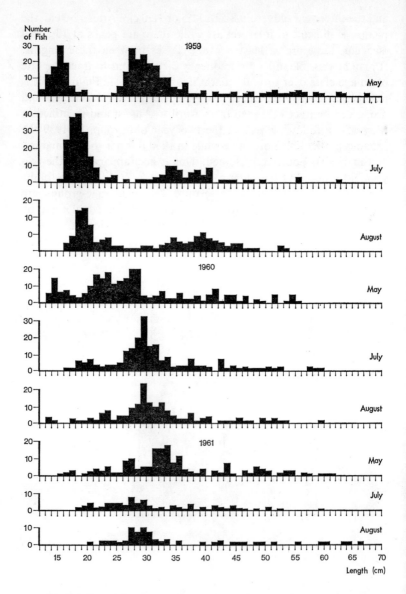

Fig. 8.4. Length frequency distribution of pike in Loch Achanalt, Loch a'Chuilinn and Loch Luichart (from: Mills, 1964).

This requires a variety of sampling techniques to insure that all age-groups in the population are sampled adequately, bearing in mind that there can be a change in behaviour and habitat with age. Population estimates are numerous and may consist of involved mathematical formulae. However, one of the most commonly used formulae and probably one of the most reliable under certain conditions is, perhaps not surprisingly, also one of the simplest. This is the Petersen estimate or Lincoln index which has been modified by Bailey.

Bailey derived the formula:

$$\hat{N} = M \frac{(n+1)}{(m+1)}$$

where $\hat{N}$ = total population size

M = number taken in first sample, marked and released

n = number taken in second sample

m = number of marked individuals recaptured in second sample.

Where it is not possible to mark individuals another method that can be used involves two successive catches c_1 and c_2 taken with the same effort from the population. An estimate of the size of the population, $\tilde{n}$, is given by:

$$\tilde{n} = c_1{}^2(c_1 - c_2)$$

with a variance

$$\text{var}[\tilde{n}] = \left[c_1^2 \, c_2^2 \, (c_1 + c_2) \right] (c_1 - c_2)^{-4}$$

The square root of this quantity is the standard error (S.E.) of the estimated number of fish. According to standard statistical techniques the approximate 95 per cent confidence limits for the number of fish in the population will be

$$\tilde{n} \pm (S.E.) \, (1\cdot96)$$

This method is more accurate in running waters or canals than in large standing waters. Although for certain species of fish such as the Miller's Thumb or Bullhead (*Cottus gobio*) it is found to give a poor estimate with rather wide confidence limits. This is because the fish hide under stones and one does not catch a large enough proportion of the total population in the first catch.

A more reliable method of estimating the population is that given by De Lury. This method of estimation requires two conditions:

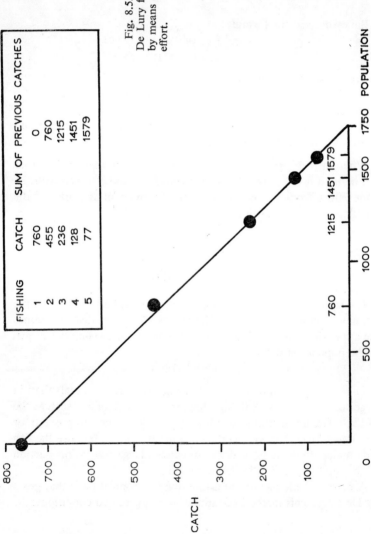

Fig. 8.5. Graphical method used by De Lury for estimation of population by means of successive units of fishing effort.

FISHING	CATCH	SUM OF PREVIOUS CATCHES
1	760	0
2	455	760
3	236	1215
4	128	1451
5	77	1579

SUM OF PREVIOUS CATCHES

POPULATION

CATCH

(1) During the successive fishings, the migration and natural mortality of the fish must be negligible.

(2) The possibility of capture must remain identical during the successive fishings, that is to say the proportion of the number of fish captured to the number of fish remaining must remain the same for each unit of fishing effort. The coefficient of capture must therefore remain invariable.

If these two conditions exist, the following formula can be deduced:

$$C = k\ (M - S)$$

in which C = catch per unit of fishing effort

M = total population present before the first unit of fishing effort

S = sum of the catches made in the course of preceding units of fishing effort

k = coefficient of capture.

If successive values of C are placed along the ordinate, and successive values of S along the abscissa, the points so obtained and joined up give a straight line, k being constant. At the point at which this line crosses the abscissa, C is equal to O and M is equal to S (Fig 8.5). Instead of making an estimate of the total number of individuals present, separate estimates for different classes of age or size can be made.

Population dispersion

It is essential to appreciate the way in which individuals in a population may be distributed, as sampling methods and statistical tests may be valid for populations having a certain distribution and not another. There are three main patterns: (*a*) random, (*b*) uniform, which is more regular than random, and (*c*) clumped, being irregular and non-random. In clumped populations such as shoals of fish, one might obtain a sample which will give either too high or too low an estimate of density depending on whether one sampled a whole shoal or only part of one.

Population regulation

Populations are regulated by density-dependent and density-independent mechanisms. Processes which are density-dependent cause a

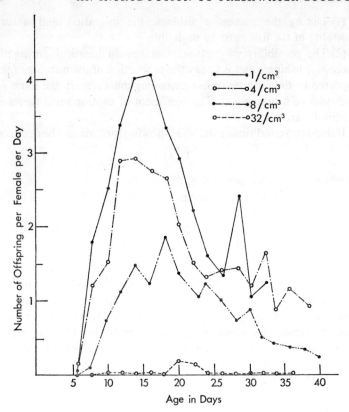

Fig. 8.6. Average number of births per female per day in the water flea, *Daphnia pulex*, at different densities (from: Frank, Boll and Kelly, 1957).

population to increase when density is low and to decrease when density is high; i.e. there are high death-rates or low growth-rates at high levels of population density and the opposite at low densities. In a study of the water flea (*Daphnia pulex*) Frank, Boll and Kelly (1957) demonstrated the effect of crowding on birth-rate, with birth-rate showing a consistent decline with increased crowding. (Fig. 8.6).

The regulation of populations of young trout and salmon has also been found to be controlled by a density-dependent mechanism – territorial behaviour. It has been found that if more than a certain number of trout or salmon alevins are stocked in a given area the survival rate decreases (Fig. 8.7). Trout and young salmon are very aggressive fish and establish territories on the stream-bed which they defend against intruders (Fig. 8.8*a*)*b*). When all available territories in a length of stream are occupied the surplus fry must emigrate, or disperse in search of vacant territories, or starve. Density independent mechanisms include climatic (e.g. rainfall, drought) and environmental (e.g. pollution) factors.

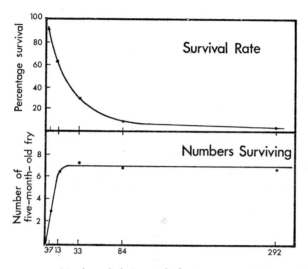

Number of alevins stocked per square metre

Fig. 8.7. The survival of brown trout fry when stocked at different densities. Upper graph: The percentage surviving from starting to feed (beginning of May) until the beginning of September. Lower graph: The actual numbers per square metre recovered in September; both plotted against the numbers of alevins stocked per square metre (after Le Cren, 1961, reproduced by permission of the Association of River Authorities).

Suggested exercises

In the following exercises rather extensive field work is undertaken which involves equipment such as a sweep net, grab and even electric fishing apparatus. These are expensive pieces of equipment and therefore may not be readily available to many. A small-meshed sweep net of approximately 40 metres in length is usually made to order by net manufacturers and may cost up to £150. However, for small streams a useful net can be made up of fine mesh netting nailed at each end to a broom pole. The recommended depth of netting is about two metres and the length can be decided to suit local conditions. Before attaching the netting to the poles, barrel leads are threaded on to a rope which is bound on to the bottom of the net and net corks threaded on to the rope to be attached to the top of the net. The completed net can be pushed forward through the water by two people holding on to the broom handles. The best methods for successful operation will come with experience, but for good results it is essential to work the net from deep into shallow water making sure the footrope is on the bottom and not lifting the net up until the river-bank is reached. A gill net can be used to sample fish in a pond. This is a sheet of netting (20 metres long by 2 metres deep) with corks along the top and lead weights along the bottom. The net is set out and anchored at each end. Netting sheets of various mesh (38 to 76 mm) may be used. An electric fishing apparatus is not difficult to construct and the pupils in the physics department might consider making one as part of a project. Circuit diagrams are described in some of the scientific literature which is well reviewed in *Fishing with Electricity* published by arrangement with FAO by Fishing News (Books) Ltd. Most electric fishing machines are powered by 12-volt car batteries or generators. Electric fishing is an illegal method of taking fish and permission for its use must be obtained from the Secretary of State in Scotland and the Minister of Agriculture, Fisheries and Food in England. Information on where grabs can be purchased can be had from the Freshwater Biological Association, Windermere Laboratory, The Ferry House, Ambleside. However, the construction of one might be undertaken by the metal-working section.

1. Make a census of the fish populations in the stream, canal or pond you have under study. Measure (from the tip of the nose to the fork of the tail) and weigh (on a spring or pan balance) all fish caught. Very small fish can be weighed in the field in bulk in a polythene bag. A sample of scales should be taken from each fish from just above the lateral line in front of the dorsal fin. Keep a sample of fish of various lengths to determine their sex and for analysis of stomach contents. Measure the area of water sampled in order to determine the density and standing crop of fish. Con-

struct a table similar to Table 10 and make a length/frequency diagram for each species of fish (see Figs 8.3 and 8.4). When estimating the total populations of the various fish species, use both methods given in this chapter.

2. Examine the scales of the fish caught and construct growth curves. Having obtained a knowledge of the mean length (and length range) of fish at the end of each year of life, plot a graph showing the percentage occurrence of the various age-classes in the population.

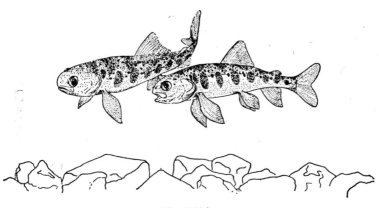

Fig. 8.8(a)

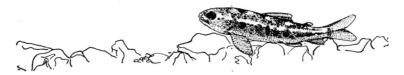

Fig. 8.8(b)

Fig. 8.8. Territorial behaviour of young salmon. (a) The fish on the right is in frontal display of moderate intensity, as it approaches a laterally-displaying fish. (b) The fish resting on the bottom shows coloration and fin positions of the typical submissive individual. The other fish is typical of an active territory-holder. Note the difference in the eyes (from: Keenleyside and Yamamoto, 1962).

3. Examine the stomach contents of the fish using identification keys for the various animals. Make a table with columns for percentage frequency of occurrence for each food organism, number of each type of organism per stomach and volume (in cm³) of each type of organism. This is done by measuring the volume of alcohol displaced by the organisms in a graduated cylinder.

4. Take grab samples across transects at right angles to the river- or canal-bank. Make a profile diagram of the river at the site of transect and note the nature of the substrate and presence of weeds and bank vegetation. Count and weigh the organisms in each grab sample and note the total biomass and calorific value for each species. The calorific value (in kcals) can be obtained from two scientific articles: (1) by M. Stube entitled *The fauna of a regulated lake*, published in 1958 in the *Annual Report of the Institute of Freshwater Research, Drottningholm*, Sweden, **39**, pp. 162–224, and (2) by K. W. Cummins and J. C. Wirycheck, entitled *Calorie equivalents for investigations in ecological energetics*, published in 1971, in *Communications of the International Association of Applied Limnology*, **98**.

5. Grow culture of *Daphnia pulex* in small aquaria and record the decrease in the number of young produced as the population increases. *Daphnia* can be found in most duck-ponds and can be fed in aquaria on *Chlorella*, which can be grown in a pure culture. A solution of yeast can also be used, but be careful not to add too much, as this quickly de-oxygenates the water. You can investigate this relationship using correlation or regression analysis. Transforming your data to logs will improve the results. These experiments are described very fully in *The vital statistics of laboratory cultures of Daphnia pulex DeGeer as related to density*, by Frank, Boll and Kelly in *Physiological Zoology*, 1957, **30**, 287–305, and '*Growth, size and reproduction in Daphnia (Crustacea: Cladocera)*. by Green in *Proceedings of the Zoological Society of London*, 1956, **126**: 173–204.

6. Study the behaviour of young trout or salmon in your aquarium, and note the various types of display. (Refer to Keenleyside, M. H. A. and Yamomoto, F. T., 1962. *Territorial behaviour of juvenile Atlantic salmon* in *Behaviour*, **19** (1): 139–169.) Note the number of attacks by individual fish on others in 5 minutes before, during and after feeding. Do any fish appear to be dominant?

9   Measurement of Production

The production of organic matter is carried out by plants through the process of photosynthesis. So the measurement of primary production resolves itself to the measurement of the photosynthetic rates of aquatic plant communities. In the case of large rooted plants the most direct method of measuring production is by the increase or decrease in standing crop over a measured interval of time. It may be expressed as the increase in (*a*) volume of plants (*b*) displacement volume (*c*) wet or damp weight (*d*) dry weight or (*e*) ash-free dry weight. This method is quite unsuitable for measuring the production of the phytoplankton since changes in the biomass reflect the net change resulting both from production, removal by the grazing of invertebrate animals, water movements and sinking. In this case it is necessary to measure as nearly as possible the instantaneous production rate. This means confining measurements of photosynthesis to a few hours. By using the formula

$$CO_2 + H_2O \rightarrow CH_2O + O_2$$

one can equate one atom of carbon, reduced to organic carbon, to one molecule of oxygen.

The most recently developed experimental method of determining organic production is that of measuring the rate of uptake of radioactive carbon (C^{14}) by the plants. The method consists of adding a small quantity of $C^{14}O_3^{\equiv}$ to a sample of water containing its natural plankton population. After allowing photosynthesis to proceed for a suitable period, the plankton is filtered on filter paper capable of retaining all autotrophic organisms. The filter is then washed, dried, and its activity measured by standard counting techniques.

A further method is to relate the chlorophyll of aquatic plants to their total organic matter and, by following the change in the chlorophyll content of natural waters, thereby estimating production. Chlorophyll 'a' is normally the most abundant and important pigment in living material and it is this which is estimated.

The gross primary production (i.e. the amount of carbon assimilated by photosynthesis) over a length of stream can be measured by monitoring, continuously or at frequent intervals, the oxygen content of the water entering and leaving it and applying the formula:

$$Q = P - R + D + A$$

where Q = rate of change of dissolved oxygen, P = rate of gross primary production, R = rate of respiration, D = rate of uptake from (or loss to) the atmosphere, and A = rate of drainage accrual, all in terms of oxygen per unit area or volume. It is possible to work out P by working on short time intervals throughout the period of twenty-four hours and assuming that R is constant and so can be determined at night, that A can be measured or is negligible and that D is determinable.

The annual production of animals (secondary production) is the actual amount of new living matter of the kind under consideration which has been produced during the year, either by the growth of old individuals, or by the production and subsequent growth of new individuals, whether these individuals have survived to the end of the year or not. Knowledge of how much animal tissue is produced in a year in a given area is most valuable. Figures for standing crop only tell us what weight of organisms are present in a given area at that time, they do not tell us what weight of animals has been produced in a year in that area. A comparison of the productivity of streams or lakes must therefore be based on production estimates and not on figures for standing crop. For example, two streams may have the same standing crop but one stream may contain a dense population of old but small, slow-growing trout while the other has a population of young, fast-growing trout. In the former the annual production is low in relation to standing crop while in the latter production is high. The evaluation of the productivity of fish in fresh waters is fundamental for establishing a rational plan for the exploitation of waters. It is also essential for attacking problems which arise from the effects of pollution on fish populations.

The annual production of stream invertebrates can be estimated by determining the numbers and weight of the organisms present in a certain area each month. By adding all the increases which occur during the year one arrives at a minimum estimate of production. Decreases can be regarded as being due to losses downstream and death.

To estimate production by a population of fish it is necessary to measure the changes in population size and growth during the period of production. There are a number of ways in which fish production can be estimated. Perhaps the most satisfactory and the simplest is the graphical method. In this, the number of individuals (N) in the population at successive time instants is plotted against the mean weight ($\bar{w}$) of the individual at the same instants. With data for a single age group over twelve months one can then plot the weight survival curve (Fig. 9.1), which is known as an Allen curve. By counting the squares on the graph paper, one can measure the area beneath the curve which is equivalent to the annual production. The inset enlargement of the Allen curve in Fig. 9.1 includes a period of negative production. Production in October is the area between the curve and weight axis or about 0·8 kg. Because no growth occurred in November, there is no area to add for production in that month. In December and January, weight loss occurred and the area (production) beneath the curve in these months should be subtracted. No production took place in February. Subsequently tissue was elaborated by the population and the area beneath the curve for March and April should be added.

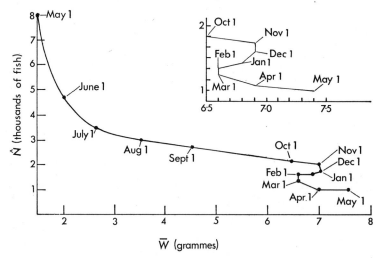

Fig. 9.1. Graphical estimation of production for a population from data for a single age group over 12 months (from: Chapman, 1968).

Where it is not possible to collect the data required for the above method of estimating annual production approximate estimates can be made using the formula:

$$K = B \times k \times 10$$

Where K = annual production in kg/ha; B = 'biogenic capacity' and k = coefficient of productivity. 'Biogenic capacity' is evaluated on a scale from 1 to 10. Factors contributing to high 'biogenic capacity' are: abundance of rooted aquatic plants, abundance of invertebrate fish food, reasonably high temperatures and adequate dissolved oxygen. 1, 2, 3 correspond to poor waters, 4, 5, 6 to average waters and 7, 8, 9 and 10 correspond to rich waters. The coefficient of productivity is composed of four secondary coefficients k_1, k_2, k_3 and k_4. The coefficient k_1 is given a value according to average annual water temperature; k_2 according to acidity or alkalinity of the water, and k_3 according to the type of fish (salmonids or cyprinids) (Table 11). Using this method in running waters the standing crop is usually found to be equal to twice the productivity as calculated by the formula.

TABLE 11

The coefficient of productivity.

k_1 = annual average temperature	k_2 = acidity or alkalinity of the water	k_3 = type of fish
10° C $k_1 = 1 \cdot 0$	acid waters	salmonids
16° C $k_1 = 2 \cdot 0$	$k_2 = 1 \cdot 0$	$k_3 = 1 \cdot 0$
22° C $k_1 = 3 \cdot 0$	alkaline waters	running water cyprinids
28° C $k_1 = 4 \cdot 0$	$k_2 = 1 \cdot 5$	$k_3 = 1 \cdot 5$
		still water cyprinids
		$k_3 = 2 \cdot 0$

(from: Huet, 1964)

Suggested exercises

1. Measure the annual production of stream invertebrates, using the method described in this chapter.
2. Measure the annual production of the fish populations in the stream under study using the graphical method. It is realized that the time required for this project might make it impossible to carry out in the time allotted and most probably most pupils will find it more convenient to carry out the method in exercise 3.
3. Make approximate estimates of the annual production of streams and rivers in your area, basing your estimates on the formula:

$$K = B \times k \times 10$$

10 Pollution

So far we have only looked at the freshwater environment in its natural state when the water is pure. Sometimes this state is upset by man when he allows poisonous substances to enter rivers in order to be diluted and carried away. These substances are in liquid form and when discharged into the river are known as effluents. Effluents are the waste products of industrial processes or domestic activities and contaminate the water. Such contamination is known as pollution.

Pollution of the water by these effluents is mainly due to poisons, suspended matter or de-oxygenation. Some effluents, such as sewage, may produce all three types of pollution.

Poisons

Poisons in solution occur in the effluents of many industries. They include acids and alkalis; chromium salts from tanneries and electro-plating works; phenols and cyanides from chemical industries and mines, and insecticides from sheep dips and agricultural chemicals. The commonest poisonous inorganic substances are chlorine, ammonia and hydrogen sulphide and the salts of many heavy metals, such as copper, lead, zinc, chromium and mercury. Very small amounts of lead, zinc and copper can remove all animal life from streams.

The insecticides most toxic to freshwater life are the organo-chlorines such as DDT and *dieldrin*. These are the chemicals in the fly sprays used in our homes. They are also used in sheep dips and for spraying crops and forests and it is during the course of such uses that they enter the water and kill insect and fish life.

Suspended matter

If the suspended matter is light or very finely divided it does not settle quickly but makes the river opaque or cloudy. This murkiness in the water prevents the penetration of sunlight and so inhibits plant

83

growth. This type of suspended matter is found in the waste water from china clay works, coal-washing effluents and mine water. In addition mine waters and drainage waters from opencast coal sites have iron in them and this precipitates on the stream-bed as orange iron ochre or ferric hydroxide.

When the particles of suspended matter are large the deposits smother all algal growth, kill rooted plants and mosses and alter the nature of the bottom. The silt from gravel-washing plants may change the nature of the stream-bed. The coarser sand particles from these gravel-washing plants may have a scrubbing effect on the rocks and gravel of the stream. This will remove the algae and bottom-living animals. Coarser rock particles may plug up the spaces in the gravel and so reduce the habitat of the bottom-living animals and smother the eggs of salmon and trout lying in the gravel. Heavy concentrations of suspended solids may block the gills of fish and cause asphyxiation.

De-oxygenation

De-oxygenation is usually caused by the decomposition of organic matter by bacteria. Organic matter is present in the effluents from a wide range of sources including domestic sewage, silage, manure heaps and cattleyards, slaughter houses, food-processing plants and paper mills.

The polluting effect of organic matter is related to the amount of oxygen taken up by bacteria in bringing about the decomposition of the material. If large quantities of dissolved oxygen in the water are used up by the bacteria as a result of this decomposition the water becomes devoid of oxygen and this results in foul smells and the death of all forms of plant and animal life. A large amount of oxygen is used up in the breakdown of sewage effluent and the oxygen concentration in the water can drop below the necessary amount required by fish. At high water temperatures this type of pollution may be more serious than at low water temperatures. This is because, as we have seen in Chapter 2, there is less dissolved oxygen in warm water than in cold water and because fish need more oxygen at high water temperatures.

We can estimate the intensity of this type of pollution by determining the amount of oxygen taken up by 1 litre of the effluent at 20 °C in five days. This intensity is called the *Biochemical Oxygen Demand*, or B.O.D. for short. Some effluents, such as silage and pig manure, have a very high oxygen demand and must be diluted many times before they are harmless to aquatic life. Rivers are to some extent self-purifying and they are thus capable of receiving a certain quantity of waste matter without any harm resulting. If the discharge is small, biological action in the presence of dissolved oxygen soon changes the waste into harmless material, but when the quantity of liquid waste is large in relation to the volume of water into which it is discharged the natural purifying process cannot cope and serious pollution results.

We can tell how bad pollution by organic matter is by the physical and chemical characteristics of the water and the type of animals present (Fig. 10.1). The first animals to disappear in polluted water are the nymphs of the stoneflies and mayflies. As the pollution becomes more severe the caddis larvae disappear and also fish such as trout, grayling, chub and dace which prefer oxygen-rich water. As pollution becomes very severe the shrimp, water louse, leech and snail disappear and also fish species such as roach and gudgeon which are more tolerant of low oxygen concentrations. When the pollution is so bad that there is little oxygen in the water the only animals present are the red chironomids or bloodworms and the oligochaete worm *Tubifex* (Tables 12 and 13). It is therefore possible to see how polluted a river is and how far downstream the effects of pollution extend by sampling the bottom fauna at the point at which the effluent is discharged and at intervals of 100 metres or so downstream of the discharge point. We can make quantitative measurements in the same way as we did when sampling the bottom fauna in a healthy stream.

If pollution is to be prevented in situations where the natural purification process cannot deal adequately with the polluting effluent, resort must be made to artificial means of purification. In some instances domestic sewage and industrial wastes can best be treated separately before discharge but it is generally more economical to collect them both together in the public sewage system and convey them to a central plant for treatment. The degree of purification required is governed by the size of the river and the use to which the river water downstream of the discharge is likely to be put. For example treated sewage effluent being discharged into a river from which water is abstracted for human consumption lower down must be of a much higher standard than that being discharged to a tidal estuary where the dilution factor is high.

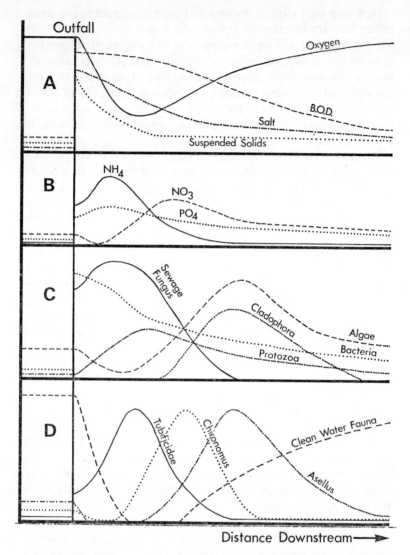

Fig. 10.1. Diagrammatic presentation of the effects of an organic effluent on a river and the changes as one passes downstream from the outfall. A and B physical and chemical changes, C changes in microorganisms, D changes in larger animals (from: Hynes, 1960).

There are various ways of disposing of or treating effluents depending upon the industry and the nature of the waste. Effluents may be disposed to rivers, sewers, tidal waters, to the land or to disused mines or mine shafts. The various forms of treatment include biological treatment, chemical precipitation, screening, lagooning, settling, floatation, spreading on the land, aeration and filtration.

TABLE 12

Class of Stream	Prominent Animals	Average Chemical Analyses in parts per million		Approximate 'Use' Classification
		B.O.D. (5 day)	D.O. (summer)	
I	Trout Grayling Stoneflies Mayflies	0 to 3	10 to 8	Domestic Water Supply
II	Club Dace Caddis Shrimps	3 to 10	9 to 7	Agriculture, Industrial processes and wash waters.
III	Roach Gudgeon Hog-lice (Asellus) and Leeches	10 to 15	7 to 5	Condenser Water Irrigation.
IV	No fish Red Chironomids (bloodworms)	15 to 30	5 to 2	Very little. Unsuitable for any amenity use
V	Barren or with fungus or small worms (tubifex)	over 30	below 2	None.

(Courtesy of the Trent River Authority.)

Biological treatment Biological treatment of sewage effluent is simply an artificial intensification and acceleration of the ordinary processes of natural purification that go on in rivers polluted by limited amounts of organic wastes. The modern biological methods of sewage treatment are (1) *Activated sludge*, which incorporates digesters to treat the sludge. The methane gas produced is frequently used to supply power at generating stations. (2) *Biological percolating*

TABLE 13

Classification of rivers in accordance with their visible degree of cleanness, based on riverside observations under normal summer conditions. (From 8th Report of Royal Commission on Sewage Disposal, Vol. II, Sect. 6, Cd. 6943, H.M.S.O. 1913)

Observed condition of river water	Very clean	Clean	Fairly clean	Doubtful	Bad
Suspended matter	Clear	Clear	Fairly clear	Slightly turbid	Turbid
Opalescence	Bright	Bright	Slightly opalescent	Opalescent	Opalescent
Smell on being shaken in bottle	Odourless	Faint earthy smell	Pronounced earthy smell	Strong earthy or wormy smell	Soapy faecal or putrid smell
Appearance in bulk	Limpid	—	Slightly brown and opalescent	Black looking	Brown or black and soapy looking
Delicate fish	May be plentiful	Scarce	Probably absent	Absent	Absent
Coarse fish	—	Plentiful	Plentiful	Scarce	Absent
Stones in shallows	Clean and bare	Clean	Lightly coated with brown fluffy deposit	Coated with brown fluffy deposit	Coated with grey growth and deposit
Stones in pools	Clean and bare	Covered with fine light brown deposit	Lightly coated with brown fluffy deposit	Coated with brown fluffy deposit	Coated with brown or black mud
Water weeds	Scarce	Plentiful. Fronds clean except in late autumn	Plentiful. Fronds brown coloured in places	Plentiful and covered with fluffy deposit	Scarce
Green algae	Scarce	Moderate quantities in shallows	Plentiful in shallows	Abundant	Abundant in protected pools
Grey algae	—	—	—	Present	Plentiful
Insects, larvae, etc.	—	—	—	Plentiful in green algae	Abundant in green alga

filtration system, where the sewage effluent, after primary sedimentation to remove large solid materials, is filtered through biologically active beds of filtering medium before final discharge.

Settling The usual method for removing settleable solids by gravitation is by the use of lagoons or settling tanks. These are particularly useful for waste waters from sand, gravel- and coal-washing plants which require little treatment other than settlement. They are also used in the paper industry to remove pulp wastes, in phosphate reduction plants and in meat and poultry processing.

Floatation   Floatation is a method used to get rid of suspended solids which do not settle easily, and also oil, grease and fat. This process involves causing the matter to rise to the surface of a liquid as a floating sludge, usually by aeration.

The River Purification Boards in Scotland and the River Authorities in England have powers to lay down standards in respect of the nature, composition, temperature, volume and rate of discharge of all effluents to all rivers. Those in respect of publicly owned sewage purification works effluents are usually determined by reference to the recommendations contained in the Eighth Report of the Royal Commission on Sewage Disposal published in 1913.

This Report recommends that an effluent should have a Bio-chemical Oxygen Demand of not more than 20 parts per million and a suspended solids content of not more than 30 parts per million. If the effluent on entering the watercourse is not diluted at least eight times, then a correspondingly lower B.O.D. and suspended solids content is required.

Suggested exercises

1. If you have a river in your area, list the sites at which effluents are discharged and note the types of effluents. Plot the sites on a map of the river. Find out from local angling clubs if there have been any cases of fish mortalities due to pollution and note their frequency and nature.

    Data on discharge points will be available from your local River Authority (there is one in nearly every county in England and Wales e.g. Yorkshire Ouse and Hull River Authority, Lancashire River Authority) or local River Purification Board (there are nine in Scotland, south of Inverness). The public health officer is also likely to have this information. These authorities are usually most co-operative and will do all they can to help. Their published annual reports give chemical and biological data on most polluted waters in the area.

2. If there is any pollution carry out a bottom fauna survey above and below the site of pollution. By looking at the numbers of

individuals of each species taken above and below the pollution source, you can test to see if significant differences exist using a χ^2 test on the contingency table. Collect water samples in polythene bottles and have them analysed for B.O.D., suspended solids, pH and alkalinity, chloride, ammoniacal nitrogen, nitrous nitrogen and nitric nitrogen. One method for determining B.O.D. is that given by McLusky (1971).

The following should be noted when the water samples are taken (a) the place where the sample was taken, (b) the temperature of the water or effluent, (c) the time of taking the sample, (d) the state of tide if an estuary. Also write down a brief description of the state of the river and banks, the height of water and its colour and any smell present.

For most tests samples can be taken twice – once during the week and once at the weekend (i.e. when the offending polluter *may* have stopped discharging effluent). A check at night might also be rewarding. Sampling can be repeated when there is a change in the water level of the river, (i.e. during drought, normal flow and flood conditions). Tests for B.O.D. can be taken more frequently, as values for this can vary considerably over a very short space of time depending on river-flow and amount and condition and nature of the effluent.

3. Arrange a visit to (a) your local sewage works (b) any local industry to see how they deal with their effluents (c) the laboratories of your local River Authority or local River Purification Board.

4. Arrange to hire the film *The River Must Live* produced by Shell. This is a very good film and deals with river pollution and the biochemical cycle.

5. Two very good sets of film strip, together with notes, on *Water Pollution* have been produced by Diana Wyllie Ltd., 3 Park Road, Baker Street, London N.W.1.

Postscript

Those of you who have been able to carry out some or all of the suggested exercises will have accumulated quite a lot of data. In addition to the data being of value in future work it could also be of great interest and value to other freshwater ecologists and conservationists. Universities, schools or individuals may therefore consider joining associations or societies concerned with this subject. They can then attend meetings, at which ideas and information can be exchanged, and receive publications dealing with the freshwater environment. Three of the most important organizations in this field in the United Kingdom are the *British Ecological Society*, the *Freshwater Biological Association* and the *Field Studies Council*.

Results which are written up in a concise and scientific manner are always favourably considered for publication in a number of scientific journals such as *Freshwater Biology, Journal of Ecology, Journal of Animal Ecology* and *Journal of Fish Biology*. The addresses of the editors of these journals are given below. The results of surveys not meeting the more rigid requirements of scientific journals can usually be published in the Transactions and Proceedings of local natural history societies.

Addresses:

H. C. Gilson,
Director,
Freshwater Biological Association,
Windermere Laboratory, Far Sawrey,
Ambleside, Westmorland.

M. J. Chadwick,
Hon. Secretary,
British Ecological Society,
Department of Biology,
The University,
Heslington, York.

The Secretary,
Field Studies Council,
9 Devereux Court,
London W.C.2.

Freshwater Biology,
Dr. T. T. Macan,
Freshwater Biological Association,
Windermere Laboratory,
Far Sawrey, Ambleside, Westmorland.

Journal of Ecology,
Prof. G. F. Asprey,
Dept. of Botany,
University College,
P.O. Box 78, Cardiff CF1 1XC.

Journal of Animal Ecology,
H. N. Southern,
Dept. of Zoology,
Oxford. (For papers on vertebrates.)

Dr. E. Broadhead,
Dept. of Zoology,
The University,
Leeds, 2. (For papers on invertebrates.)

Journal of Fish Biology,
L. E. Mawdesley-Thomas,
Huntingdon Research Centre,
Huntingdon, PE18 6ES.

References

In addition to the scientific works referred to in the text, and listed below, a number of additional books are included as suggestions for further reading; these are marked with an asterisk.

BACKIEL, T. and LE CREN, E. D. 1967. 'Some density relationships for fish populations, in *Biological Basis of Freshwater Fish Production*, Gerking, S. D. (Ed.), Blackwell.

*CARPENTER, K. E. 1928. *Life in Inland Waters*, Sidgwick and Jackson.

*CLEGG, J. 1952. *The Freshwater Life of the British Isles*, Warne.

*CLEGG, J. 1956. *The Observer's Book of Pond Life*, Warne.

*COKER, R. E. 1954. *Streams, Lakes and Ponds*, University of North Carolina Press.

EGGLISHAW, H. J. 1968. 'The quantitative relationship between bottom fauna and plant detritus in streams of different calcium concentrations. *J. appl. Ecol.*, 5, 731–740.

EGGLISHAW, H. J. and MORGAN, N. C. 1965. 'A survey of the bottom fauna of streams in the Scottish Highlands. Part II. The relationship of fauna to the chemical and geological conditions.' *Hydrobiologia*, XXVI (1–2), 173–183.

ELLIOTT, J. M. 1971. *Some Methods for the Statistical Analysis of Samples of Benthic Invertebrates*, Freshwater Biological Association, Scientific Publication No. 25.

ENGELHARDT, W. 1964. *Pond Life* (Young Specialist Series), Burke.

FRANK, P., BOLL, C. D. and KELLY, R. W., 1957. 'The vital statistics of laboratory cultures of *Daphnia pulex* De Geer as related to density.' *Physiol. Zool.*, 30: 287–305. University of Chicago Press.

*GOLTERMAN, H. L. and CLYMO, R. S. (Ed.). 1960. *Methods for the Chemical Analysis of Fresh Waters*, Blackwell.

GREEN, J. 1956. 'Growth, size and reproduction in *Daphnia* (*Crustacea: Cladocera*)'. *Proc. Zool. Soc.*, 126: 173–204.

HERRINGTON, R. B. and DUNHAM, D. K. 1967. 'A technique for sampling general fish habitat characteristics of streams.' U.S. Dep. Agr., Forest Serv., Intermountain Forest and Range Exp. Sta., Ogden, Utah.

HUET, M. 1964. The evaluation of fish productivity in fresh waters. *Verh. Internat. Verein. Limnol.*, 15, 524–8.

*HYNES, H. B. N. 1960. *The Biology of Polluted Waters*, University of Liverpool Press.

*HYNES, H. B. N. 1971. *The Ecology of Running Waters*, University of Liverpool Press.

*JENKINS, J. T. 1925. *The Fishes of the British Isles*, Warne.

KEENLEYSIDE, M. H. A. and YAMAMOTO, F. T. 1962. Territorial behaviour of juvenile Atlantic salmon. *Behaviour*, **19**(1), 139–169.

KNIGHT, C. B. 1965. *Basic Concepts of Ecology*, Macmillan.

KORMONDY, E. J. 1969. *Concepts of Ecology*. Prentice-Hall.

LE CREN, E. D. 1961. *How many fish survive? Yearbook of the River Boards Association*, **9**, 57–64.

*MACAN, T. T. 1959. *A Guide to Freshwater Invertebrate Animals*, Longman.

MACAN, T. T. 1963. *Freshwater Ecology*, Longman.

MACAN, T. T. 1970. *Biological Studies of the English Lakes*, Longman.

MACAN, T. T. and WORTHINGTON, E. D. 1951. *Life in Lakes and Rivers*, Collins.

MCCLINTOCK, D. and FITTER, R. S. R. 1956. *Pocket Guide to Wild Flowers*, Collins.

MCLUSKY, D. S. 1971. *Ecology of Estuaries*. Heinemann.

MANN, K. H. 1964 a. 'The pattern of energy flow in the fish and invertebrate fauna of the River Thames.' *Proc. 1st British Coarse Fish Conf.*, 58–61.

MANN, K. H. 1964b. 'The pattern of energy flow in the fish and invertebrate fauna of the River Thames.' *Verh. Internat. Verein. Limnol.*, XV, 485–495.

*MARTIN, W. K. 1965. *The Concise Flora in Colour*, Joseph.

*MELLANBY, H. 1938. *Animal Life in Freshwater*, Methuen.

MILL, H. R. 1895. 'Bathymetrical survey of the English Lakes.' *Geogr. J.*, **6**, 47–73; 135–166.

MILLS, D. H. 1964. 'The ecology of the young stages of the Atlantic salmon in the River Bran, Ross-shire.' *Freshwat. Salm. Fish. Res.*, **32**, 58 pp.

MILLS, D. H. 1970. Preliminary observations on fish populations in some Tweed tributaries. *Annual Report to the Tweed Commissioners, 1970*, Appendix III, 16–22.

MILLS, D. H. 1971a. *Salmon and Trout: A Resource its Ecology, Conservation and Management,* Oliver and Boyd.

MILLS, D. H. 1971b. 'The growth and population densities of roach in some Scottish waters.' *Proc. 4th British Coarse Fish Conf.*

MITCHELL, I. 1971. 'Duddingston Loch. A study in limnology and eutrophication problems.' Honours thesis for the degree of B.Sc. (Ecological Science), University of Edinburgh.

MORGAN, N. C. and WADDELL, A. 1961. 'Insect emergence from a small trout loch and its bearing on the food supply of fish.' *Freshwat. Salm. Fish. Res.*, **25**.

*MUIRHEAD-THOMSON, R. C., 1971. *Pesticides and the Freshwater Fauna*, Academic Press.

MURRAY, J. and PULLAR, L. 1910. *Bathymetrical Survey of the Scottish Freshwater Lochs*, Vol. 1, Challenger Office.

*MUUS, B. J. and DAHLSTROM, P. 1971. *Collins Guide to the Freshwater Fishes of Britain and Europe,* Collins.

NIELSEN, A. 1950. 'The torrential invertebrate fauna.' *Oikos,* **2,** 176–196.

ODUM, E. 1959. *Fundamentals of Ecology,* Saunders.

ODUM, E. 1963. *Ecology,* Holt, Rinehart and Winston.

*PHILLIPSON, J. 1966. *Ecological Energetics,* Studies in Biology No. 1, Arnold.

*POPHAM, E. J. 1955. *Some Aspects of Life in Fresh Water,* Heinemann.

*REID, G. M. 1961. *Ecology of Inland Waters and Estuaries,* Reinhold.

REYNOLDSON, T. B. 1958. 'Observations on the comparative ecology of lake-dwelling triclads in southern Sweden, Finland and northern Britain.' *Hydrobiologia,* **12,** 129–141.

RICKER, W. E. (Ed.) 1968. *Methods for Assessment of Fish Production in Fresh Waters,* Blackwell.

RUSSELL-HUNTER, W. 1970. *Aquatic Productivity,* Collier-Macmillan.

*RUTTNER, F. 1953. *Fundamentals of Limnology,* University of Toronto Press.

SCOTT, D. 1958. 'Ecological studies on the Trichoptera of the River Dean, Cheshire.' *Arch. Hydrobiol.,* **54,** 340–392.

STUBE, M. 1958. 'The fauna of a regulated lake.' *Rep. Inst. Freshw. Res. Drottning.,* **39,** 55–98.

TANSLEY, A. G. 1939. *The British Islands and their Vegetation,* Cambridge University Press.

VALLENTYNE, J. R. 1967. 'A simplified model for instructional use.' *J. Fish. Res. Bd. Canada,* **24** (11): 2473–2479.

VARLEY, M. E. 1967. *British Freshwater Fishes,* Fishing News (Books) Ltd.

*VOLLENWEIDER, R. A. 1969. *A Manual on Methods for Measuring Primary Production in Aquatic Environments,* Blackwell.

*WELCH, P. S. 1948. *Limnological Methods,* McGraw-Hill.

WELCH, P. S. 1952. *Limnology* (2nd edition), McGraw-Hill.

*WHITTAKER, R. H. 1970. *Communities and Ecosystems,* Macmillan.

Index

Notes